Triumph Herald Owners Workshop Manual

by J. L. S. Maclay

Models covered:

948 c.c. Triumph Herald Saloon	- April 1959 to March 1961
948 c.c. Triumph Herald Coupe	- April 1959 to June 1961
948 c.c. Triumph Herald Convertible	- March 1960 to June 1961
948 c.c. Triumph Herald 'S' Saloon	- February 1961 to January 1964
1,147 c.c. Triumph Herald 12 Saloon	- April 1961 on
1,147 c.c. Triumph Herald 1200 Coupe	- April 1961 to October 1964
1,147 c.c. Triumph Herald 1200 Convertible	- April 1961 to September 1967
1,147 c.c. Triumph Herald 1200 Estate	- May 1961 to September 1967
1,147 c.c. Triumph Herald 1200 Van	- February 1962 to October 1964
1,147 c.c. Triumph Herald 12/50 Saloon	- March 1963 to September 1967
1,296 c.c. Triumph Herald 13/60 Saloon	- October 1967 on
1,296 c.c. Triumph Herald 13/60 Convertible	- October 1967 on
1,296 c.c. Triumph Herald 13/60 Estate	- October 1967 on

ISBN 900550 10 4

© Haynes Group Limited 1971, 1978, 1986

All rights reserved. No part of this book may be reproduced or transmitted in any form or by any means, electronic or mechanical, including photocopying, recording or by any information storage or retrieval system, without permission in writing from the copyright holder.

(8M1)

Haynes Group Limited
Haynes North America, Inc

www.haynes.com

Disclaimer

There are risks associated with automotive repairs. The ability to make repairs depends on the individual's skill, experience and proper tools. Individuals should act with due care and acknowledge and assume the risk of performing automotive repairs.

The purpose of this manual is to provide comprehensive, useful and accessible automotive repair information, to help you get the best value from your vehicle. However, this manual is not a substitute for a professional certified technician or mechanic.

This repair manual is produced by a third party and is not associated with an individual vehicle manufacturer. If there is any doubt or discrepancy between this manual and the owner's manual or the factory service manual, please refer to the factory service manual or seek assistance from a professional certified technician or mechanic.

Even though we have prepared this manual with extreme care and every attempt is made to ensure that the information in this manual is correct, neither the publisher nor the author can accept responsibility for loss, damage or injury caused by any errors in, or omissions from, the information given.

ACKNOWLEDGEMENTS

My thanks are due to British Leyland for the generous assistance given in the supply of technical material and illustrations; to Castrol Ltd, for supplying the lubrication chart; to the Champion Sparking Plug Company who supplied the illustrations showing the various spark plug conditions; and to the Editor of Autocar for permission to use the cutaway drawing featured on the cover. The bodywork repair photographs used in this manual were provided by Lloyds Industries Limited who supply 'Turtle Wax', 'Duplicolor Holts', and other Holts range products. Special thanks are due to Mr. L. Tooze and Mr. R.T. Grainger whose experience and practical help were of great assistance in the compilation of photographs for this manual.

Although every care has been taken to ensure all data in this manual is correct, bearing in mind that manufacturers' current practice is to make small alterations and design changes without reclassifying the model, no liability can be accepted for damage, loss or injury caused by any errors or omissions in the information given.

PHOTOGRAPHIC CAPTIONS & CROSS REFERENCES

For ease of reference this book is divided into numbered chapters, sections and paragraphs. The title of each chapter is self explanatory. The sections comprise the main headings within the chapter. The paragraphs appear within each section.

The captions to the majority of photographs are given within the paragraphs of the relevant section to avoid repetition. These photographs bear the same number as the sections and paragraphs to which they refer. The photograph always appears in the same Chapter as its paragraph. For example if looking through Chapter Ten it is wished to find the caption for photograph 9:4 refer to section 9 and then read paragraph 4.

To avoid repetition once a procedure has been described it is not normally repeated. If it is necessary to refer to a procedure already given this is done by quoting the original Chapter, section and sometimes paragraph number.

The reference is given thus: Chapter No. /Section No. Paragraph No. For example Chapter 2, section 6 would be given as: Chapter 2/6. Chapter 2, Section 6, paragraph 5 would be given as Chapter 2/6:5. If more than one section is involved the reference would be written: Chapter 2/6 to 7 or where the section is not consecutive 2/6 and 9. To refer to several paragraphs within a section the reference is given thus: Chapter 2/6. 2 and 4.

To refer to a section within the same Chapter the Chapter number is usually dropped. Thus if a reference in a Chapter 4 merely reads 'see section 8', this refers to section 8 in that same Chapter.

All references to components on the right or left-hand side are made as if looking forward to the bonnet from the rear of the car.

HERALD 1200 ESTATE

HERALD 13/60 SALOON

948 c.c. HERALD SALOON

HERALD 1200 CONVERTIBLE

Contents

Acknowledgements	2
Photographic captions and cross references	3
Introduction	6
Routine maintenance	7
Recommended lubricants	10
Lubrication chart	11
Ordering spare parts	12
Chapter 1 Engine	13
Chapter 2 Cooling system	54
Chapter 3 Fuel system and carburation	62
Chapter 4 Ignition system	84
Chapter 5 Clutch and actuating mechanism	94
Chapter 6 Gearbox	104
Chapter 7 Propeller shaft and universal joints	118
Chapter 8 Rear axle	122
Chapter 9 Braking system	128
Chapter 10 Electrical system	143
Chapter 11 Suspension, dampers and steering	169
Chapter 12 Bodywork and underframe	185
Safety first!	202
Index	203

INTRODUCTION

This is a manual for do-it-yourself minded Triumph Herald Owners. It shows how to maintain these cars in first class condition and how to carry out repairs when components become worn or break. Regular and careful maintenance is essential if maximum reliability and minimum wear are to be achieved.

The step-by-step photographs show how to deal with the major components and in conjunction with the text and exploded illustrations should make all the work quite clear — even to the novice who has never previously attempted to more complex job.

Although Heralds are hardwearing and robust it is inevitable that their reliability and performance will decrease as they become older. Repairs and general reconditioning will be come necessary if the car is to remain roadworthy. Early models requiring attention are frequently bought by the more impecunious motorist who can least afford the repair prices charged in garages, even though these prices are usually quite fair bearing in mind overheads and the high cost of capital equipment and skilled labour.

It is in these circumstances that this manual will prove to be of maximum assistance, as it is the ONLY workshop manual written from practical experience specially to help Herald owners.

Manufacturer's official manuals are usually splendid publications which contain a wealth of technical information. Because they are issued primarily to help the manufacturers authorised dealers and distributors they tend to be written in very technical language, and tend to skip details of certain jobs which are common knowledge to garage mechanics. Owner's workshop manuals are different as they are intended primarily to help the owner. They therefore go into many of the jobs in great detail with extensive photographic support to ensure everything is properly understood so that the repair is done correctly.

Owners who intend to do thier own maintenance and repairs should have a reasonably comprehensive tool kit. Some jobs require special service tools, but in many instances it is possible to get round their use with a little care and ingenuity. For example a 3½ in. diameter jubilee clip makes a most efficient and cheap piston ring compressor.

Throughout this manual ingenious ways of avoiding the use of special equipment and tools are shown. In some cases the proper tool must be used. When this is the case a description of the tool and its correct use is included.

When a component malfunctions repairs are becoming more and more a case of replacing the defective item with an exchange rebuilt unit. This is excellent practice when a component is thoroughly worn out, but it is a waste of good money when overall the component is only half worn, and requires the replacement of but a single small item to effect a complete repair. As an example, a non-functioning dynamo can frequently be repaired quite satisfactory just by fitting new brushes.

A further function of this manual is to show the owner how to examine malfunctioning parts; determine what is wrong, and then how to make the repair.

Given the time, mechanical do-it-yourself aptitude, and a reasonable collection of tools, this manual will show the ordinary private owner how to maintain and repair his car really economically.

ROUTINE MAINTENANCE

The maintenance instructions listed below are basically those recommended by the manufacturer. They are supplemented by additional maintenance tasks which, through practical experience, the author recommends should be carried out at the intervals suggested.

The additional tasks are indicated by an asterisk and are primarily of a preventive nature in that they will assist in eliminating the unexpected failure of a component due to fair wear and tear.

The levels of the engine oil, radiator cooling water, windscreen washer water and battery electrolyte, also the tyre pressures, should be checked weekly or more frequently if experience dictates this to be necessary. Similarly it is wise to check the level of the fluids in the clutch and brake master cylinder reservoirs at monthly intervals. If not checked at home it is advantageous to use regularly the same garage for this work as they will get to know your preferences for particular oils and the pressures at which you like to run your tyres.

6,000 miles

EVERY 6,000 MILES (or six months if 6,000 miles is not exceeded).

1. Run the engine until it is hot and place a container of at least 8 pints capacity under the drain plug on the left-hand side of the sump and allow the oil to drain for at least 10 minutes. Clean the plug and the area around the plug hole in the sump and replace the plug, tightening it firmly. Clean the oil filler with petrol, refill the sump with 7 pints of a recommended grade of oil (see page 10) and clean off any oil which may have been spilt over the engine or its components. The interval between oil changes should be reduced in very hot or dusty conditions or during cold weather with much slow stop/start driving.
2. Check the valve clearances and adjust, if necessary, as described on page 48.
3. Check and adjust, if necessary, the engine slow running as described on pages 71, 72 and 78.
4. Check and adjust the brakes if necessary. The procedure is described on page 129.
5. Examine and renew any defective hoses in the braking system. Ensure that there is adequate clearance between them and any chassis or other components to eliminate chafing.
6. Examine the tyres and should wear be apparent take the appropriate action to correct the cause e.g. mis-alignment, poor balancing, over or under inflation. If in any doubt a Triumph repair garage should be consulted especially where alignment is suspect because complicated and expensive equipment is required to carry out the necessary check. Remove any flints or other road matter from the treads.
7. Apply grease (see page 10) to the handbrake cable guides and the compensator sector.
8. Lubricate with a recommended grade of oil (page 10) all hinges, catches and controls to allow them to work freely and to prevent unnecessary wear.
9. Remove the plug on the lower steering swivels (one on each side) and fit a screwed grease nipple. Jack up the front road wheels and apply a grease gun filled with a recommended HYPOID oil (see page 10) and pump until oil exudes from the swivel.
10. Remove the sparking plugs for cleaning and reset the gaps to .025 in. (.64 mm.). Clean the ceramic insulators and examine them for cracks or other damage likely to cause 'tracking'. Test the plugs before refitting and renew any which are suspect.
11. Release the spring clips and remove the distributor cap and rotor arm. Apply a few drops of thin oil (page 10) over the screw in the centre of the cam spindle and on the moving contact breaker pivot. Grease the cam surface very lightly. Remove any excess oil or grease with a clean rag. Apply a few drops of oil through the hole in the contact breaker base plate to lubricate the automatic timing control.
12. Clean and adjust the contact breaker points as described on page 85.
13. The various types of air cleaner are described on page 63 and these should be cleaned, refilled with oil or have the paper element renewed as appropriate. In very dusty conditions the intervals for carrying out this task may well have to be reduced considerably.
14. The fan belt must be tight enough to drive the generator without overloading the generator and water pump bearings. The method of adjusting the fan belt is described on page 58 and is correct when it can be pressed inwards $\frac{3}{4}$ in. (19 mm.) on the longest run - from the generator pulley to the crankshaft pulley.
15. Check the operation of all electrical equipment particularly stop/tail lamps, plate illumination, and side lamps. Adjust, if necessary, the headlamp settings.
16. On engines fitted with S.U. carburetters remove the hexagon caps from the dashpot/s and top them up to within $\frac{1}{2}$ in. of the top with SAE 20 oil (see page 10). Remove the float chambers, empty away any sediment, check the condition of the needle valve, clean and reassemble. Remove the filter in the carburetters if fitted.
17. On pre-1965 models, with the car standing on level ground, check the level of oil in the gearbox by removing the oil filler/level plug on the right-hand side of the gearbox. Top up, if necessary with SAE 90 EP gear oil (page 10) until the oil starts to run out of the filler hole.
18. Give the bodywork and chromium trim a thoroughly good wax polish. If a chromium cleaner is used to remove rust on any part of the car's plated parts remember that the cleaner also removes part of the chromium so use sparingly.
19. Remove the carpets or mats and thoroughly vacuum clean the interior of the car. Beat out or vacuum clean the carpets. If the upholstery is soiled apply an upholstery cleaner with a damp sponge and wipe off with a clean dry cloth.
20. Hoods and tonneau covers should be cleaned with

ROUTINE MAINTENANCE

plain soap and water. Detergents, caustic soaps or spirit cleaners should never be used.

12,000 miles

EVERY 12,000 MILES (or every 12 months if 12,000 miles is not exceeded).

1. Carry out all the maintenance tasks listed for the 6,000 miles service.
2. On post 1965 models, with the car standing on level ground, remove the oil level plug at the right-hand side of the gearbox and, using a pump type oil can with a flexible nozzle fitted with a HYPOID oil (page 10), top up the gearbox until the oil is level with the bottom of the filler plug threads. Allow any surplus oil to drain away, clean the area around the plug holes, and the plug itself, and replace the plug tightening it firmly. On pre 1965 models the gearbox must be drained and refilled with the recommended grade of oil.
3. Remove the oil plug at the rear top of the rear axle casing and, using a pump type oil can with a flexible nozzle filled with a HYPOID oil (page 10), top up the rear axle until the oil is level with the bottom of the filler plug threads. Allow any surplus oil to drain away, clean the area around the plug hole, and the plug itself, and replace the plug tightening it firmly.
4. Unscrew the oil filter from the left-hand side of the engine and discard it. Fit a new one, ensuring an oil tight seal by cleaning and smearing the joint faces with oil before screwing the oil filter unit tightly home.
5. Unscrew the bolt on the top cover of the fuel pump and remove them both. Lift out the filter gauze from its seating and wash it in petrol. Loosen the sediment with the aid of a small screwdriver and blow it out with compressed air (a foot operated tyre pump is ideal for this purpose). It is important not to disturb or damage the non-return valve during this operation. Renew the cork gasket if it is hardened or broken. Assemble the filter gauze into its seating, taking care to place the gauze face downwards so that it can be removed easily when required. Do not forget to refit the fibre washer under the head of the dome cover retaining bolt and tighten it just enough to make a leak-proof joint.
6. Jack up the car and remove the road wheels and brake drums. Remove the dust from the drums and clean the back plates. Examine the brake shoes and renew worn or contaminated shoes (further details on page 130). Reassemble and adjust, (page 129). In the case of disc brakes examine the pads for wear and deterioration (see page 136).
7. Examine and, if necessary, tighten the front and rear suspension attachments, steering connections, water pump, starter motor generator and generator pulley, oil filter and all universal joints and bolts.
8. Inject a few drops of engine oil (page 10) through the hole in the rear of the generator.
9. It is wisdom to renew the sparking plugs every 12,000 miles.
10. Remove the plug from the top of the steering unit and fit a screwed grease nipple. Apply a grease gun filled with a recommended grease (page 10) and give it five strokes only. Remove the nipple and

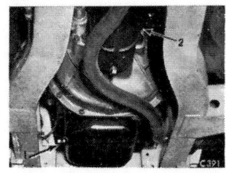

The sump drain plug and gearbox filler plug.

The rear axle filler plug.

Replacing the oil filter.

1. Screw
2. Washer
3. Dome cover
4. Gasket
5. Gauze
6. Valve

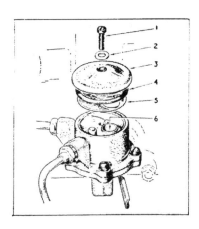

ROUTINE MAINTENANCE

refit the plug. Over greasing can cause damage to the rubber bellows.

11. The front hubs must be repacked with grease (page 10). Full details are given on page 138.

12. Remove the plug on each rear hub and fit a screwed grease nipple. Apply a grease gun filled with grease (page 10) until grease exudes from the bearing.

14. Unscrew the plug from the top of the water pump and fit a screwed grease nipple. Apply the grease gun filled with a recommended grease (page 10) giving it five strokes only. Remove the nipple and refit the plug.

* 14. It is a sound scheme to visit your local main agent and have the underside of the body steam cleaned. This will take about 1½ hours and cost about £4. All traces of dirt and oil will be removed and the underside can then be inspected carefully for rust, damaged hydraulic pipes, frayed electrical wiring and similar maladies. The car should be greased on completion of this job.

* 15. At the same time the engine compartment should be cleaned in the same manner. If steam cleaning facilities are not available then brush 'Gunk' or a similar cleaner over the whole engine and engine compartment with a stiff paint brush working it well in where there is an accumulation of oil and dirt. Do not paint the ignition system but protect it with oily rags when the Gunk is washed off; as the Gunk is washed away it will take with it all traces of oil and dirt, leaving the engine looking clean and bright.

24,000 miles

EVERY 24,000 MILES (or every two years).

1. Carry out the maintenance tasks listed for the 6,000 and 12,000 miles services.

2. On post 1965 cars drain and refill with the appro-
* priate grades of oil (page 10) the gearbox and rear axle. This is recommended so that any minute particles of metal are carried away in the old oil so helping to prevent further wear.

The steering unit greasing plug.

The plug for greasing rear hubs.

The water pump greasing plug.

Steering lower swivel lubrication.

The cable guides and compensator sector.

RECOMMENDED LUBRICANTS

BRITISH ISLES (ALL SEASONS)

Component	Mobil	Shell	Esso	B.P.	Castrol	Duckham's	Regent
ENGINE	Mobiloil Special	Shell Super Oil	Esso Extra Motor Oil 20W/30	Energol Motor Oil 20W or Visco Static or Visco Static Long Life	Castrolite	Duckham's No1 Twenty or Duckham's Q5500	Havoline 20/20W or Special 10W/30 Havoline
KING PIN LOWER SWIVEL, GEARBOX, REAR AXLE	Mobilube GX.90	Shell Spirax 90 E.P.	Esso Gear Oil GP.90/140	Energol SAE.90 EP	Castrol Hypoy	Duckham's Hypoid 90	Universal Thuban 90
FRONT & REAR HUBS, STEERING UNIT, ENGINE WATER PUMP	Mobilgrease M.P.	Shell Retinax A	Esso Multi-Purpose Grease H	Energrease L2	Castrolease L.M.	Duckham's L.B.10	Marfak Multi-Purpose 2
OIL CAN	Mobil Handy Oil	Shell X-100 20W	Engine Oil	Energol Motor Oil SAE 20W	Everyman Oil	Duckham's General Purpose Oil	Havoline 20/20W

OVERSEAS COUNTRIES

Component		Mobil	Shell	Esso	B.P.	Castrol	Duckham's	Caltex Texaco	S.A.E. & A.P.I. Designation
ENGINE Air Temp. °F.	Over 80°	Mobiloil Special	Shell Super Oil	Esso Motor Oil 30 / ESSO EXTRA MOTOR OIL 10W/30	Energol SAE 30 / VISCO STATIC	Castrol X.L.	Duckham's No1 Thirty / Q5500	Havoline 30 / HAVOLINE SPECIAL 10W/30	S.A.E. 30 M.M.
	30° — 80°			Esso Motor Oil 20	Energol Motor Oil SAE 20W	Castrolite	Duckham's No1 Twenty	Havoline 20/20W	S.A.E. 20 M.M.
	Below 30°			Esso Motor Oil 10W	Energol Motor Oil SAE 10W	Castrol Z	Duckham's No1 Ten	Havoline 10W	S.A.E. 10 M.M.
KING PIN LOWER SWIVEL, GEARBOX, REAR AXLE	Over 30°	Mobilube GX.90	Shell Spirax 90 EP.	Esso Gear Oil GP. 90	Energol SAE.90EP	Castrol Hypoy	Duckham's Hypoid 90	Universal Thuban 90	G.L.4 Hypoid 90
	Below 30°	Mobilube GX.80	Shell Spirax 80 EP.	Esso Gear Oil GP. 80	Energol SAE.80EP	Castrol Hypoy Light	Duckham's Hypoid 80	Universal Thuban 80	G.L.4 Hypoid 80
FRONT & REAR HUBS, STEERING UNIT, ENGINE WATER PUMP		Mobilgrease MP.	Shell Retinax A	Esso Multi-Purpose Grease H	Energrease L2	Castrolease L.M.	Duckham's LB.10	Marfak Multi-Purpose 2	

LUBRICATION CHART

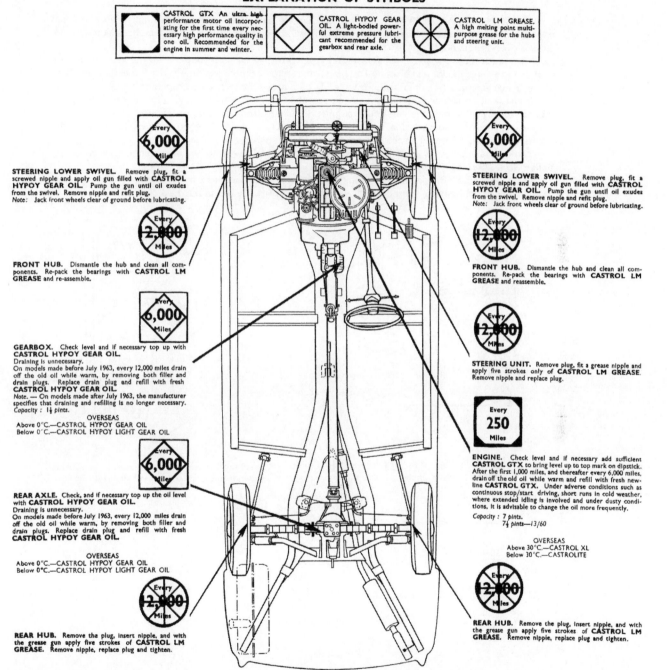

ORDERING SPARE PARTS

Always order genuine British Leyland spare parts from your nearest Triumph dealer or local garage. Authorised dealers carry a comprehensive stock of GENUINE PARTS and can supply most items 'over the counter'.

When ordering new parts it is essential to give full details of your car to the storeman. He will want to know model and chassis numbers, and in the case of engine spares the engine number. Year of manufacture is helpful too. If possible take along the part to be replaced.

If you want to retouch the paintwork you can obtain an exact match (providing the original paint has not faded) by quoting the paint code number in conjunction with the model number.

The chassis or commission number as it is called at the factory is stamped on a model identification plate located on the left-hand side of the scuttle.

The engine number is stamped on a flat surface on the left-hand side of the engine immediately under No. 4 sparking plug.

When obtaining new parts remember that many assemblies can be exchanged. This is very much cheaper than buying them outright and throwing away the old part.

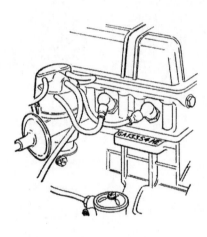

The engine number is stamped on a flange on the left-hand side of the cylinder block.

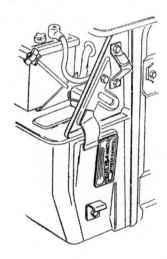

The chassis, or commission, number is stamped on a plate attached to the left-hand side of the scuttle.

CHAPTER ONE

ENGINE

CONTENTS

General Description	1
Routine Maintenance	2
Major Operations with Engine in Place	3
Major Operations with Engine Removed	4
Methods of Engine Removal	5
Engine Removal, without Gearbox	6
Engine Removal with Gearbox	7
Dismantling the Engine - General	8
Removing Ancilliary Engine Components	9
Cylinder Head Removal - Engine on Bench	10
Cylinder Head Removal - Engine in Car	11
Valve Removal	12
Valve Guide Removal	13
Dismantling the Rocker Assembly	14
Timing Cover, Gears & Chain Removal	15
Camshaft Removal	16
Distributor Drive Removal	17
Sump, Piston, Connecting Rod & Big End Bearing Removal	18
Gudgeon Pin - Removal	19
Piston Ring Removal	20
Flywheel & Engine Endplate Removal	21
Crankshaft & Main Bearing Removal	22
Lubrication & Crankcase Ventilation Systems - Description	23
Oil Filter - Removal & Replacement	24
Oil Pressure Relief Valve - Removal & Replacement	25
Oil Pump - Removal & Dismantling	26
Timing Chain Tensioner - Removal & Replacement	27
Examination & Renovation - General	28
Crankshaft Examination & Renovation	29
Big End & Main Bearings - Examination & Renovation	30
Cylinder Bores - Examination & Renovation	31
Pistons & Piston Rings - Examination & Renovation	32
Camshaft & Camshaft Bearings - Examination & Renovation	33
Valves & Valve Seats - Examination & Renovation	34
Timing Gears & Chain - Examination & Renovation	35
Timing Chain Tensioner - Examination & Renovation	36
Rockers & Rocker Shaft - Examination & Renovation	37
Tappets - Examination & Renovation	38
Flywheel Starter Ring - Examination & Renovation	39
Oil pump - Examination & Renovation	40
Cylinder Head - Decarbonisation	41
Valve Guides - Examination & Renovation	42
Sump - Examination & Renovation	43
Engine Reassembly - General	44
Crankshaft Replacement	45
Piston & Connecting Rod Reassembly	46
Piston Ring Replacement	47
Piston Replacement	48
Connecting Rod to Crankshaft Reassembly	49
Front Endplate - Reassembly	50
Camshaft Replacement	51

CHAPTER ONE

Timing Gears - Chain Tensioner - Cover Replacement...	52
Oil Pump Replacement	53
Crankshaft Rear Seal - Housing - Endplate & Flywheel Replacement	54
Sump Replacement	55
Valve & Valve Spring Reassembly	56
Rocker Shaft & Tappet Reassembly	57
Cylinder Head Replacement	58
Rocker Arm/Valve Adjustment	59
Distributor & Distributor Drive Replacement...	60
Final Assembly	61
Engine Replacement	62

The original engine fitted to the Herald was of 948 c.c. With the introduction of the Herald 1200 the engine capacity was increased to 1147 c.c. This was uprated to 1296 c.c. with the appearance of the Herald 13/60.

ENGINE SPECIFICATION & DATA - 948 c.c.

Engine - General

Type	4 cylinder in line O.H.V. pushrod operated
Bore	2.48 in. (63 mm.).
Stroke	2.992 in. (76.0 mm.).
Cubic capacity	948 c.c. (57.8 cu. in.)
Compression ratio:	
Saloon (single carb) - High C	8 to 1
- Low C	7 to 1
Coupe (twin carb) - High C	8.5 to 1
- Low C	7.4 to 1
Compression pressure:	
single carb - High C	125 lb/sq.in. (8.79 kg/cm^2)
twin carb - High C	128 lb/sq.in. (9.0 kg/cm^2)
Torque:	
single carb - High C	48 lb/ft. at 2,750 r.p.m.
twin carb - High C	49 lb/ft. at 3,000 r.p.m.
Max. B.H.P.:	
single carb - High C	34.5 at 4,500 r.p.m.
twin carb - High C	45 at 5,800 r.p.m.
Firing order	1, 3, 4, 2
Location of No.1 cylinder	Next to radiator

Camshaft

Camshaft drive	From crankshaft by single roller chain
Camshaft journal diameter...	1.8402 to 1.8407 in. (46.74 to 46.75 mm.)
Diametrical clearance	.0026 to .0046 in. (.07 to .12 mm.)
End float	.0025 to .0065 in. (.06 to .16 mm.)
End thrust	Taken by core plug

Connecting Rods & Big & Little End Bearings

Type	Angular split big end. Fully floating small end
Big end bearings - Type	Shell
Big end bearing - Internal diameter	1.626 to 1.627 in. (41.30 to 41.32 mm.)
End float on crankpin	.0086 to .0125 in. (.23 to .32 mm.)
Undersizes available	- .010, - .020, - .030, - .040 in.
	(- .254, - .508, - .762, - 1.016 mm.)
Internal diameter of small end bush	.7515 to .7525 in. (19.09 to 19.11 mm.)

Crankshaft & Main Bearings

Main journal diameter	2.0005 to 2.001 in. (50.81 to 50.83 mm.)
Crankpin diameter	1.6250 to 1.6255in. (41.27 to 41.28 mm.)
Crankshaft end thrust	Taken by thrust washers on rear main bearing
End float	.004 to .006 in.
Main bearing internal diameter...	2.0015 to 2.0025 in. (50.84 to 50.86 mm.)
Main bearing housing internal diameter ...	2.146 to 2.1465 in. (54.51 to 54.52 mm.)
Rear journal width	1.360 to 1.361 in. (34.54 to 34.57 mm.)
Thickness of thrust washers	.091 to .093 in. (2.31 to 2.36 mm.)
Oversize thrust washers	.096 to .098 in. (2.44 to 2.49 mm.)
Undersize bearings available	- .010, - .020, - .030, - .040 in.

ENGINE

Undersize bearings available	(− .254, − .508, − .762, − 1.016 mm.)

Cylinder Block
Type	Cylinder cast integral with top half of crankcase
Water jackets	Full length
Oversize bores − First	.010 in. (.254 mm.)
− Max.	.030 in. (.762 mm.)

Cylinder Head
Type	Cast iron with vertical valves
Port arrangements	Inlet and exhaust ports on same side
Number of ports − Exhaust	4 separate
− Inlet	2 siamised

Gudgeon Pin
Type	Fully floating
Fit in piston	Light push fit at 68°F (20°C)
Outer diameter	.62485 to .62510 in. (15.87 to 15.88 mm.)

Lubrication System
Type	Pressure and splash. Wet sump
Oil filter	Purolator, A.C. Delco or Tecalemit by-pass filter
Sump capacity	7 pints
Oil pump − Type	Eccentric rotor
Normal oil pressure at 2,000 r.p.m.	40 to 60 lb/sq. in. (2.8 to 4.2 kg/cm^2)

Pistons
Type	Aluminium alloy − split skirt
Number of rings	3. 2 compression, 1 oil control
Clearance in cylinder − Top	.002 to .0027 in. (.0508 to .06858 mm.)
− Bottom	.0012 to .0019 in. (.03048 to .04826 mm.)
Width of ring grooves − Compression rings	.0797 to .0807 in. (2.02 to 2.05 mm.)
− Oil control ring	.157 to .158 in. (3.99 to 4.01 mm.)
Piston oversizes available	+ .010, + .020, + .030 in.
	(+ 254, + .508, + .762 mm.)

Piston Rings
Compression rings	Parallel − chromium plated
Compression ring width	.0777 to .0787 in. (1.97 to 1.99 mm.)
Fitted gap	.007 to .012 in. (.18 to .30 mm.)
Oil control ring	Slotted scraper
Oil control ring width	.1553 to .1563 in. (3.94 to 3.97 mm.)
Oil control ring fitted gap	.007 to .012 in. (.18 to .30 mm.)

Tappets
Type	Barrel with flat base
Outside diameter	.6867 to .6871 in. (17.45 to 17.46 mm.)

Rocker Gear
Diameter of rocker shaft	.5607 to .5612 in. (14.24 to 14.26 mm.)
Bore of rockers	.562 to .563 in. (14.27 to 14.30 mm.)

Valves
Head diameter − Inlet	1.179 to 1.183 in. (29.94 to 30.05 mm.)
− Exhaust	1.058 to 1.084 in. (26.87 to 27.53 mm.)
Stem diameter − Inlet	.310 to .311 in. (7.87 to 7.89 mm.)
− Exhaust	.308 to .309 in. (7.82 to 7.85 mm.)
Stem to guide clearance − Inlet	.001 to .003 in. (.03 to .08 mm.)
− Exhaust	.003 to .005 in. (.08 to .13 mm.)
Valve stem to rocker arm clearance	.010 in. (.25 mm.)

Valve Guides
Length (single carb)	2.44 in. (61.98 mm.)
(twin carb)	2.25 in. (57.15 mm.)
Outside diameter − Inlet and exhaust	.501 to .502 in. (12.72 to 12.75 mm.)
Inside diameter − Inlet and exhaust	.312 to .313 in. (7.92 to 7.95 mm.)

CHAPTER ONE

Fitted height above head
- Inlet and exhaust ... .749 to .751 in. (19.025 to 19.075 mm.)

Valve Timing
- Inlet opens ... ... 10° B.T.D.C.
- Inlet closes ... ... 50° A.B.D.C.
- Exhaust opens ... ... 50° B.B.D.C.
- Exhaust closes ... ... 10° A.T.D.C.

Valve Springs
- Type (single carb).. ... ... Single valve spring
- (twin carb) ... ... Double inner and outer valve springs
- Fitted length - Outer ... ... 1.36 in. (34.54 mm.)
- Inner ... ... 1.14 in. (28.96 mm.)
- Fitted load - Outer ... ... 24 to 27 lbs (11 to 12 kg.)
- Inner ... ... 10 lbs (4.54 kg.)
- Number of coils - Outer 7¼
- Inner 7½

TORQUE WRENCH SETTINGS

Big end bolts ... ... 38 to 42 lb/ft. (5.254 to 5.807 kg.m.)
Clutch attachment ... ... 18 to 20 lb/ft. (2.489 to 2.765 kg.m.)
Cylinder head nuts ... ... 42 to 46 lb/ft. (5.807 to 6.36 kg.m.)
Engine mounting bolts ... ... 18 to 20 lb/ft. (2.489 to 2.765 kg.m.)
Flywheel securing bolts ... ... 42 to 46 lb/ft. (5.807 to 6.36 kg.m.)
Fuel pump bolts ... ... 12 to 14 lb/ft. (1.659 to 1.936 kg.m.)
Main bearing bolts ... ... 50 to 55 lb/ft. (6.913 to 7.604 kg.m.)
Manifold nuts ... ... 18 to 20 lb/ft. (2.5 to 2.8 kg.m.)
Oil gallery set screws ... ... 13 to 15 lb/ft. (1.8 to 2.1 kg.m.)
Oil pump to block ... ... 6 to 8 lb/ft. (0.830 to 1.106 kg.m.)
Rocker cover nuts ... ... 1½ lb/ft. (0.105 kg.m.)
Rocker pedestals ... ... 24 to 26 lb/ft. (3.318 to 3.595 kg.m.)
Sump to crankcase ... ... 16 to 18 lb/ft. (2.212 to 2.489 kg.m.)
Timing cover attachment.. ... ... 8 to 10 lb/ft. (1.106 to 1.383 kg.m.)
Water pump to cylinder head... ... ... 18 to 20 lb/ft. (2.489 to 2.765 kg.m.)
Water outlet elbow nuts ... ... 16 to 18 lb/ft. (2.212 to 2.489 kg.m.)

ENGINE SPECIFICATION & DATA - 1147 c.c.

The engine specification and data is identical to the 948 c.c. engine except for the differences listed below.

Engine - General
Bore ... ... 2.728 in. (69.3 mm.)
Cubic capacity ... ... 1147 c.c. (70.0 cu. in.)

Compression Ratios:-   Compression Pressure
Herald 1200   Engine Nos.
From   To

GA.1 HE ... ... GA.164,889 HE) ...
GA.177,973 HE .. ... GA.178,000 HE) High C   8 to 1   131 lb/sq. in.
GB.1 HE ... ... GB.2,700 HE) ...

GA.164,890 HE .. ... GA.177,972 HE) ...
GA.178,001 HE .. ... GA.190,223 HE) High C   8 to 1   131 lb/sq. in.
GA.190,301 HE .. ... GA.190,340 HE) ...

GA.190,224 HE .. ... GA.190,300 HE) ...
GA.190,341 HE .. ... GA.235,664 HE) High C   8.5 to 1   133 lb/sq. in.
GD.110,001 HE .. ... and future) ...
GA.1 LE ... ... and future... ..Low C   6.8 to 1   121 lb/sq. in.
Herald 12/50 ... ...High C)
                                         only)   8.5 to 1   136 lb/sq. in.

Torque
Herald 1200 8 to 1 High C ... ... 60 lb/ft. at 2,250 r.p.m.

ENGINE

8.5 to 1 High C	61.5 lb/ft. at 2,500 r.p.m.
6.8 to 1 Low C	56 lb/ft. at 2,250 r.p.m.
Herald 12/50 8.5 to 1 High C	63 lb/ft. at 2,600 r.p.m.

Max. B.H.P.

Herald 1200 8 to 1 High C	39 at 4,500 r.p.m.
8.5 to 1 High C	48 at 5,200 r.p.m.
6.8 to 1 Low C	43 at 4,750 r.p.m.
Herald 12/50 8.5 to 1 High C	51 at 5,200 r.p.m.

Camshaft

End float	.004 to .008 in. (.10 to .20 mm.)

Connecting Rods & Big & Little End Bearings

End float on crankpin	.0105 to .0126 (.266 to .320 mm.)
Internal diameter of small end bush	.8122 to .8126 (20.63 to 20.64 mm.)
Undersizes available	-.010, -.020, -.030 in.
	(-.254, -.508, -.762 mm.)

Crankshaft & Main Bearings

End float	.004 to .008 in.
Main bearing internal diameter	2.0015 to 2.0037 in. (50.84 to 50.89 mm.)
Rear journal width	1.2976 to 1.2995 in. (32.95 to 33.01 mm.)

Gudgeon Pin

Outer diameter	.8123 to .8125 in. (20.63 to 20.64 mm.)

Lubrication System

Oil filter	Full flow

Pistons

Type	Aluminium alloy - split skirt up to engine Nos. GA.137,545 and GD.21,229
	Aluminium alloy - solid skirt from engine Nos. GA.137,546 and GD.21,230
Number of rings	3. 2 compression, 1 oil control
Clearance in cylinder:	
Pistons made by Automotive) - Top	.0029 to .0033 in.
Engineering Co. Ltd.) - Bottom	.0011 to .0015 in.
Pistons made by British Piston) - Top	.0163 to .0193 in.
Ring Co. Ltd.) - Bottom	.0012 to .0015 in.
Pistons made by Wellworthy - Top	.0038 to .0041 in.
- Bottom	.0012 to .0015 in.

Width of Ring Grooves

Split skirt pistons - Compression rings	.0802 to .0812 in. (2.03 to 2.06 mm.)
- Oil control ring	.157 to .158 in. (3.99 to 4.01 mm.)
Solid skirt pistons - Compression rings	.0797 to .0807 in.
- Oil control ring	.157 to .158 in. (3.99 to 4.01 mm.)

Piston Rings

2nd compression ring	Early models parallel, later models tapered
Fitted gap - all rings	.008 to .013 in. (.20 to .33 mm.)
Oil control ring - width	.154 to .156 in. (3.90 to 3.96 mm.)

Valves

Head diameter - Inlet	1.304 to 1.308 in. (33.12 to 33.22 mm.)
- Exhaust	1.148 to 1.182 in. (29.16 to 29.26 mm.)
Stem diameter - Inlet	.310 to .311 in. (7.87 to 7.89 mm.)
- Exhaust	.308 to .309 in. (7.82 to 7.85 mm.)

Valve Guides

Length	2.25 in. (57.15 mm.)

Valve Timing

Herald 1200 Engine Nos:	
GA.HE to GA.164,889 HE - Inlet opens	12° B.T.D.C.

CHAPTER ONE

GA.177,973 HE to GA.178,000 HE	-Inlet closes	52° A.B.D.C.
GB. 1 HE to GB. 2,700 HE	- Exhaust opens..	52° B.B.D.C.
	- Exhaust closes	12° A.T.D.C.
All other Herald 1200	- Inlet opens ...	18° B.T.D.C.
and Herald 12/50 engines	- Inlet closes ...	58° A.B.D.C.
	- Exhaust opens..	58° B.B.D.C.
	- Exhaust closes..	18° A.T.D.C.

Valve Springs
  Type ... ... Single valve springs
  Fitted length - Herald 1200 .. .. 1.36 in. (34.54 mm.)
              - Herald 12/50 ... ... 1.38 in. (35.03 mm.)
  Fitted load - Herald 1200... ... ... 27 to 30 lbs. (12.25 to 13.61 kgs.)
              - Herald 12/50.. ... ... 32 to 42 lbs. (14.51 to 19.05 kgs.)
  No. of coils - Herald 1200... ... ... 7 1/4
              - Herald 12/50.. 6

TORQUE WRENCH SETTINGS (where different from 948 c.c. engine)

Manifold nuts	24 to 26 lb/ft. (3.318 to 3.595 kg.m.)
Oil filter to crankcase	15 to 18 lb/ft. (2.074 to 2.489 kg.m.)
Oil gallery set screws	18 to 20 lb/ft. (2.489 to 2.765 kg.m.)
Timing cover (5/16 in. setscrew)..	14 to 16 lb/ft. (1.936 to 2.212 kg.m.)
Timing cover (5/16 in. slotted screw)	8 to 10 lb/ft. (1.106 to 1.388 kg.m.)

ENGINE SPECIFICATION & DATA 1,296 c.c.

The engine specification and data is identical to the 1147 c.c. engine except for the differences listed below.

Engine - General
  Bore ... ... 2.900 in. (73.7 mm.)
  Cubic capacity ... ... 1296 c.c. (79.2 cu.in.)
  Compression ratio - High C ... ... 8.5 to 1
                   - Low C ... ... 7.5 to 1
  Compression pressure - High C .. .. 139 lb/sq.in.
                      - Low C .. .. 127 lb/sq.in.
  Torque - High C ... ... 73 lb/ft. at 3,000 r.p.m.
        - Low C ... ... 67 lb/ft. at 3,000 r.p.m.
  Max. B.H.P. - High C.. ... ... 61 at 5,000 r.p.m.
             - Low C.. ... ... 54 at 5,200 r.p.m.

Camshaft
  Camshaft journal diameter   1.9649 to 1.9654 in. (49.91 to 49.92 mm.)
  End float   .0042 to .0085 in. (.11 to .216 mm.)

Connecting Rods & Big & Little End Bearings
  Type ... ... Angular split big end. Fully floating or interference fit small end.
  End float on crankpin ... ... .0025 to .008 in. (.063 to .218 mm.)
  Internal diameter of small end bush ... ... .8110 to .8115 in. (20.51 to 20.612 mm.)

Crankshaft & Main Bearings
  Main bearing internal diameter ... ... 2.002 to 2.0025 (50.85 to 50.86 mm.)

Cylinder Head
  Number of ports - Exhaust... ... ... 4 separate
                 - Inlet ... ... 4 separate

Gudgeon Pin
  Type ... ... Fully floating or interference fit
  Fit in piston (fully floating) .. .. Light push fit at 68°F (20°C)

Pistons
  Type ... ... Aluminium alloy - solid skirt
  Clearance in cylinder ... ...
  Pistons made by Brico Co. Ltd. - Top .. .020 to .025 in.
                                - Bottom .. .0019 to .0024 in.

ENGINE

Pistons made by Hepworth Co.Ltd. - Top	.0201 to .0248 in.
- Bottom	.0019 to .0024 in.
Width of ring grooves - Compression rings	.0797 to .0807 in.
- Oil control rings	.157 to .158 in.

Piston Rings
Oil control ring	Three piece slotted scraper
Fitted gap. Compression & oil control	.012 to .022 in. (.30 to .85 mm.)

Tappets
Outside diameter	.7996 to .8000 in. (20.294 to 20.320 mm.)

Valves
Head diameter - Exhaust	1.168 to 1.172 in. (29.66 to 29.76 mm.)
Stem diameter - Inlet	.310 to .3112 in. (7.874 to 7.90 mm.)
- Exhaust	.310 to .3105 in. (7.874 to 7.887 mm.)
Stem guide clearance - Inlet	.0008 to .0023 in. (.02 to .06 mm.)
- Exhaust	.0015 to .003 in. (.0261 to .075 mm.)

Valve Guides
Length	2.0625 in. (52.387 mm.)

Valve Timing
- Inlet open	18° B.T.D.C.
- Inlet closes	58° A.B.D.C.
- Exhaust opens	58° B.B.D.C.
- Exhaust closes	18° A.T.D.C.

TORQUE WRENCH SETTINGS (where different from 1147 c.c. engines)

Oil pump to block	8 to 10 lb/ft. (1.106 to 1.383 kg.m.)

1. GENERAL DESCRIPTION

The engine is a four-cylinder, overhead valve type. It is supported by rubber mountings in the interests of silence and lack of vibration.

Two valves per cylinder are mounted vertically in the cast iron cylinder head and run in pressed in valve guides. They are operated by rocker arms, pushrods and tappets from the camshaft which is located at the base of the cylinder bores in the left-hand side of the engine. The correct valve stem to rocker arm pad clearance can be obtained by the adjusting screws in the ends of the rocker arms.

On all except 1,296 c.c. models the cylinder head has two siamised inlet, and four separate ports on the right-hand side. 1,296 c.c. models have four inlet and four exhaust ports.

The cylinder block and the upper half of the crankcase are cast together. The bottom half of the crankcase consists of a pressed steel sump.

The pistons are made from anodised aluminium alloy with split or solid skirts. Two compression rings and a slotted oil control ring are fitted. The gudgeon pin is retained in the little end of the connecting rod by circlips or by interference fit.

Renewable shell type big end bearings are fitted.

At the front of the engine a single chain drives the camshaft via the camshaft and crankshaft chain wheels which are enclosed in a pressed steel cover.

The chain is tensioned automatically by a spring blade which presses against the non-driving side of the chain so avoiding any lash or rattle.

The camshaft is supported by four bearings bored directly into the cylinder block except on certain later engines which are fitted with special replaceable bearings. Endfloat is controlled by a forked locating plate positioned on the front end plate.

The statically and dynamically balanced forged steel crankshaft is supported by three renewable thinwall shell main bearings which are in turn supported by substantial webs which form part of the crankcase. Crankshaft endfloat is controlled by semi-circular thrust washers located on each side of the rear main bearing.

The centrifugal water pump and radiator cooling fan are driven, together with the dynamo, from the crankshaft pulley wheel by a rubber/fabric belt. The distributor is mounted in the middle of the left-hand side of the cylinder block and advances and retards the ignition timing by mechanical and vacuum means. The distributor is driven at half crankshaft speed by a short shaft and skew gear from a skew gear on the camshaft located between the second and third journals.

The oil pump is located in the crankcase and is driven by a short shaft from the skew gear on the camshaft.

Attached to the end of the crankshaft by four bolts and two dowels is the flywheel to which is bolted the Borg & Beck clutch. Attached to the engine end plate is the gearbox bellhousing.

2. ROUTINE MAINTENANCE

1. Once a week, or more frequently if necessary, remove the dipstick and check the engine oil level which should be at the 'MAX' mark. Top up the oil in the sump with the recommended grade (see page

10 for details). On no account allow the oil to fall below the 'MIN' mark on the dipstick.

2. Every 6,000 miles run the engine till it is hot; place a container with a capacity of at least 7 pints under the drain plug in the sump; undo and remove the drain plug; and allow the oil to drain for at least ten minutes. While the oil is draining wash the oil filler cap gauge in petrol, shake dry, and re-oil.

3. Clean the drain plug, ensure the washer is in place, and return the plug to the sump, tightening the plug firmly. Refill the sump with 7 pints of the recommended grade of oil (see page 10 for details). Every 12,000 miles the oil filter element should be renewed as described in Section 24.

4. In very hot or dusty conditions; in cold weather with much slow stop/start driving, with much use of the choke; or when the engine has covered a very high mileage, it is beneficial to change the engine oil every 3,000 miles, and the filter every 6,000 miles.

3. MAJOR OPERATIONS WITH ENGINE IN PLACE

The following major operations can be carried out to the engine with it in place in the body frame:-
1. Removal and replacement of the cylinder head assembly.
2. Removal and replacement of the sump.
3. Removal and replacement of the big end bearings.
4. Removal and replacement of the pistons and connecting rods.
5. Removal and replacement of the timing chain and gears and the timing cover oil seal.
6. Removal and replacement of the camshaft.
7. Removal and replacement of the oil pump.

4. MAJOR OPERATIONS WITH ENGINE REMOVED

The following major operations can be carried out with the engine out of the body frame and on the bench or floor:-
1. Removal and replacement of the main bearings.
2. Removal and replacement of the crankshaft.
3. Removal and replacement of the flywheel.

5. METHODS OF ENGINE REMOVAL

There are two methods of engine removal. The engine can either be removed complete with gearbox, or the engine can be removed without the gearbox by separation at the gearbox bellhousing. Both methods are described. Irrespective of whether the Triumph engine is removed with or without the gearbox, it will be found to be one of the easiest units to take out and replace. No pit or ramps are necessary as the jobs usually done underneath such as prop-shaft/gearbox separation are all done from inside the car. As a further bonus engine accessability is excellent.

6. ENGINE REMOVAL WITHOUT GEARBOX

1. Practical experience has proved that the engine can be removed easily in about 3 hours (less with experience) by adhering to the following sequence of operations.

2. Open the bonnet and prop it up to expose the engine and ancilliary components. Turn on the water drain taps found at the bottom of the radiator and on the side of the cylinder block. N.B. Do not

Fig. 1.1. EXPLODED VIEW OF THE CYLINDER BLOCK, CYLINDER HEAD, SUMP, AND ENGINE MOUNTINGS

1 Cylinder block. 2 Bolt and lock washer. 3 Oil gallery end plug. 4 Welch plug. 5 Welch plug for rear face and L.H. side of block. 6 Welch plug for rear of camshaft. 7 Oil gallery bolt. 8 Washer. 9 Oil pump shaft bush. 10 Cylinder head stud. 11 Cylinder head/lifting eye bracket studs. 12 Water drain plug. 13 Fibre washer. 14 Petrol pump attachment stud. 15 Distributor mounting stud. 16 Front sealing block. 17 Filler piece. 18 Screw. 19 Oil retaining cover. 20 Gasket. 21 Bolt. 22 Spring washer. 23 Oil pressure indicator switch. 24 Front engine plate. 25 Gasket. 26 Engine mounting foot for Herald 1200 models (other models differ slightly). 27 Left-hand side engine mounting foot. 28 Bolt. 29 Spring washer. 30 Nut. 31 Front engine mounting. 32 Nut. 33 Spring washer. 34 Rear engine plate. 35 Sump. 36 Gasket. 37 Oil strainer gauge. 38 Oil drain plug. 39 Bolt. 40 Dipstick. 41 Felt washer. 42 Breather pipe. 43 Deflector. 44 Bolt. 45 Nut. 46 Oil pressure relief valve piston. 47 Piston spring. 48 Retaining plug. 49 Washer. 50 Cylinder head. 51 Pushrod tubes. 52 Valve guide. 53 Core plug. 54 Water delivery tube. 55 Rocker pedestal stud. 56 Rocker cover stud. 57 Nut. 58 Gasket. 59 Inlet valve. 60 Exhaust valve. 61 Lower collar. 62 Valve spring. 63 Valve spring upper collar. 64 Tappet. 65 Pushrods. 66 Rocker shaft. 67 Rocker pedestal drilled. 68 Bolt/shakeproof washer. 69 Plain rocker pedestal. 70 Nut. 71 Lock washer. 72 Rocker No.1. 73 Rocker No.2. 74 Ball pin. 75 Locking nut. 76 Centre rocker spring (1). 77 Intermediate rocker springs (2). 78 Outer rocker springs (2). 79 Collar. 80 Mills pin. 81 Tocker cover. 82 Oil filler cap/breather. 83 Rocker cover gasket. 84 Nut. 85 Plain/fibre washers.

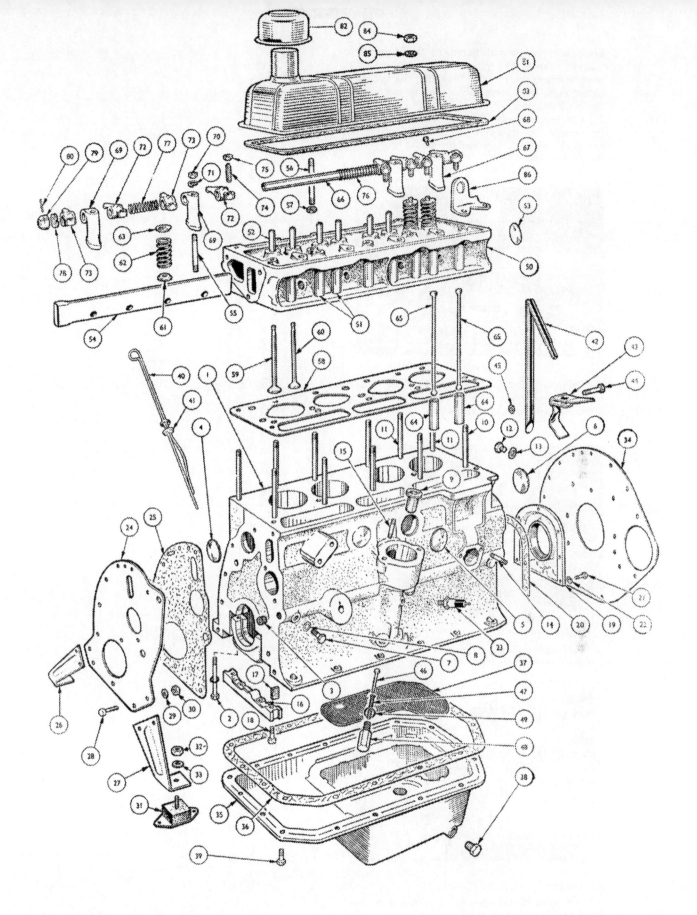

drain the water in your garage or the place where you will remove the engine if receptacles are not at hand to catch the water. Providing the bonnet is pulled right back there is no need to take it off. Drain the engine oil.

3. It is best to remove the battery completely. Undo the winged nuts (photo) which holds the battery retaining bracket in place and lift off the bracket and the side stays.

4. Undo the screws in the centre of the battery terminal caps (photo), take the terminal caps off the terminal posts, and lift out the battery.

5. Pull the small wire by its tag off the Lucar connector on the side of the distributor (photo).

6. Then undo the terminal in the centre of the coil (photo) to free the H.T. lead to the distributor cap.

Fig. 1, 2. EXPLODED VIEW OF THE MAIN MOVING ENGINE COMPONENTS

1 Crankshaft. 2 Main bearings. 3 Thrust washers. 4 Shims. 5 Crankshaft chain wheel. 6 Oil slinger. 7 Pulley wheel. 8 Woodruff key. 9 Bush for nose of gearbox input shaft. 10 Pulley wheel securing nut. 11 Flywheel. 12 Dowel. 13 Flywheel to crankshaft dowel. 14 Bolt. 15 Bolt locking plate. 16 Crankshaft. 17 Camshaft locating plate. 18 Bolt. 19 Spring washer. 20 Camshaft chain wheel. 21 Bolt. 22 Bolt locking plate. 23 Timing chain. 24 Timing chain cover. 25 Oil seal. 26 Gasket. 27 Timing chain tensioner anchor plate. 28 Rivet. 29 Spring tensioner. 30 Tensioner spring anchor pin. 31 Cotter pin. 32 Connecting rod. 33 Small end bush. 34 Hollow dowel. 35 Big end bolts. 36 Tab lock washer. 37 Big end bearings. 38 Piston. 39 Chrome plated top compression ring. 40 Plain compression ring. 41 Oil control ring. 42 Gudgeon pin. 43 Circlip. 44 Distributor and oil pump drive gear. 45 Oil pump drive shaft. 46 Gear to shaft attacking pin. 47 Distributor mounting. 48 Gasket (2 thicknesses available .006 in. and .020 in.). 49 Bolt. 50 Nut. 51 Spring washer.

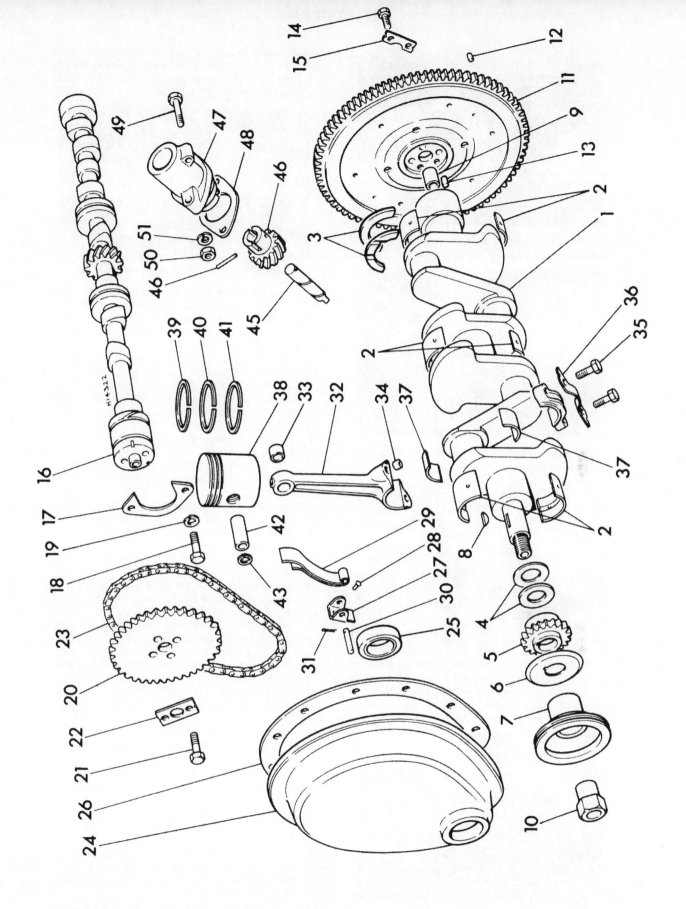

CHAPTER ONE

6.6

7. Pull the two wires by their tags off the Lucar connector on the back of the dynamo (photo).

6.7

8. Pull the wire off the connector on the oil pressure warning sender unit (photo).

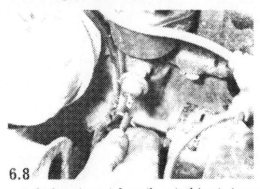

6.8

9. Undo the union nut from the petrol input pipe on the fuel pump (photo). On some models the pipe is connected by a rubber connector which is simply pulled off. Block the open end of the pipe to prevent petrol from the tank running out.

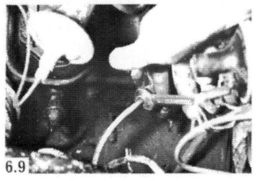

6.9

10. Different types of air cleaner have been used at various times. If the air cleaner/s gets in the way it must be removed (photo). The small pancake type can always be left in place.

6.10

11. Undo the screw from the clamp which holds the accelerator cable to the carburetter or in some instances undo the nut and washer holding the ball joint rod to the accelerator linkage arm (photo). Release the return spring/s.
12. The next stage is to release the choke cable by undoing the clamp nut (arrowed) which holds it to the linkage.
13. Pull off the wire by its tag from the Lucar terminal on the thermostat housing (photo).
14. Undo the clips on the two radiator hoses and pull the hoses off the thermostat outlet pipe, and water pump inlet pipe (photo).
15. Undo the clips on the heater inlet hoses and pull them off the inlet and outlet pipes on the rear of the engine (photo).
16. The next step is to undo the nuts and bolts from either side of the radiator. Make a special note if the horns or any other items are hung on the radiator bolts.
17. With the bolts removed the radiator can be lifted out (photo)
18. Undo the nuts from the three studs which hold the exhaust pipe to the exhaust manifold (photo).
19. Pull the exhaust pipe away from the exhaust manifold and undo the petrol overflow pipe (if fitted) from the bottom of the inlet manifold.
20. Undo the nut and bolt holding the exhaust downpipe to the clamp attached to the bellhousing bottom bolt (photo).
21. Take off the nut retaining the starter lead to the starter terminal post and pull off the starter motor lead (photo).
22. Undo the two bolts holding the starter motor in place and lift out the starter motor together with the distance piece.
23. Where fitted undo the knurled nut which holds the tachometer drive cable to the rear of the distributor (photo).
24. Pull off the advance and retard pipe from the distributor or on certain models undo the union nut which holds the pipe in place (photo).
25. Undo the nuts from the bolts and studs which hold the engine endplate to the bellhousing. Note the clip which secures the clutch hydraulic pipe to the top centre stud (photo).
26. The engine earth lead must be taken off next. It will either be attached to a bolt on the front right

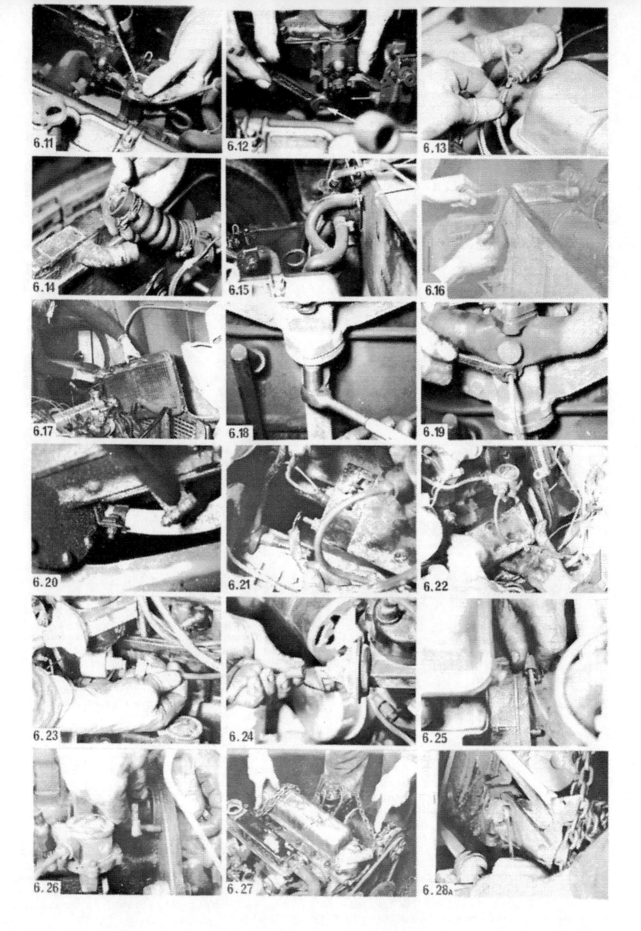

CHAPTER ONE

engine mounting bracket or to one of the bellhousing bolts (photo)

27. Attach a lifting chain or strong rope to the two lifting hook holes provided on the engine (photo). Take the weight of the engine on suitable lifting tackle. Place a jack under the gearbox to take the weight of this unit.

28. Undo the nuts which hold the front engine mountings in place. Several different types of mounting with different attachment arrangements have been used at different times. In photo 'A' two nuts have to be undone and the bolt removed, while in photo 'B' it is necessary to remove only one nut.

29. Slightly raise the engine, and the jack, and pull the engine forwards and up until the clutch is free from the first motion shaft in the gearbox. It is important that no excess load is placed on the clutch so take great care at this stage. Once clear of the bellhousing pull the engine forwards tilting the front further upwards to clear the front crossmember and wind the engine out of the car. (Photo).

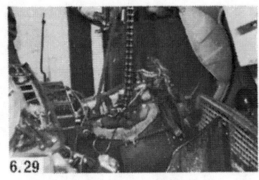

ENGINE REMOVAL WITH GEARBOX

1. Follow the instructions in Section 6 showing how to remove the engine without the gearbox omitting paragraphs 22 and 25. Do not start the work listed in paragraphs 26 to 29 until the following operations have been completed.

2. To avoid having to bleed the clutch slave cylinder on reassembly, an excellent tip worth following is to place a piece of scrap polythene under the clutch master cylinder filler cap and screw the cap down tight. (Photo). This will prevent any fluid running out of the master cylinder when the clutch pipe is disconnected.

3. Undo the union nut on the master cylinder to free the pipe which runs to the slave cylinder (photo).

4. From inside the car undo the bolts which hold the transmission cover in place (photo). NOTE On

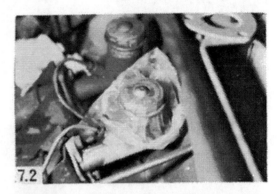

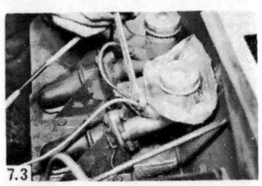

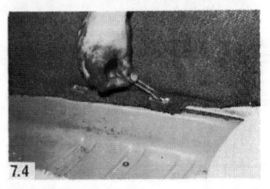

some models it will be necessary to remove the fascia support brackets first.

5. Loosen the nut under the gearchange lever ball top and unscrew the top (photo).

6. The fibreglass transmission cover can then be lifted off (photo).

7. Undo the knurled nut on the right-hand side of the gearbox extension to free the speedometer cable (Photo).

8. Undo the nut and bolt which hold the gearchange lever to the slot in the remote control shaft (photo).

9. Press down and turn the bayonet gearlever cover and lift the gearlever complete with caps and spring up out of the remote control extension (photo).

10. Carefully scribe a mating mark across the universal joint and gearbox drive flanges (arrowed) in photo).

11. Undo the four nuts and bolts which hold the universal joint to the gearbox mainshaft flange (photo).

12. Then undo the single nut and washer (photo) from each of the two gearbox extension mounting rubbers, so the rubbers can be freed from the mounting bracket.

ENGINE

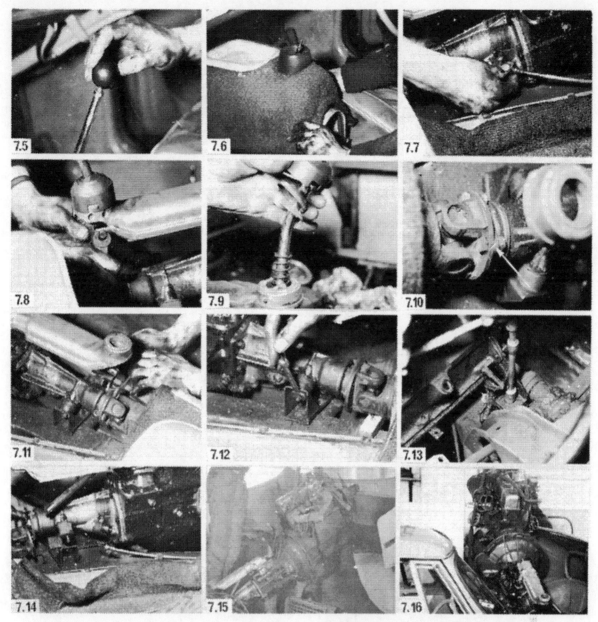

13. At the front of the engine undo the front mounting nuts. On this early model one nut has to be undone on each side (photo). See also Section 6.28.

14. Now follow the instructions in paragraphs 26 to 29, and in addition note the following points:- As soon as the engine is lifted an inch or two free the rear mounting bolts from the mounting bracket (photo).

15. The engine and gearbox should be lifted out at an angle of about 45° once the gearbox extension has cleared the chassis crossmember (photo).

16. Unless a moveable hoist is being used, it will be necessary to lift the engine about 5 feet and to then turn it through 90°. This will allow the bonnet to be closed and the car to be wheeled out from under the engine. Place a blanket over the windscreen to prevent accidental damage. To finally complete the job clear out any loose nuts and bolts from the engine compartment and place them where they will not become lost.

8. DISMANTLING THE ENGINE - GENERAL

1. It is best to mount the engine on a dismantling stand, but if one is not available, then stand the engine on a strong bench so as to be at a comfortable working height. Failing this, the engine can be stripped down on the floor.

2. During the dismantling process the greatest care should be taken to keep the exposed parts free from dirt. As an aid to achieving this, it is a sound scheme to thoroughly clean down the outside of the engine, removing all traces of oil and congealed dirt.

3. Use paraffin or a good grease solvent such as 'Gunk'. The latter compound will make the job much easier, as, after the solvent has been applied and allowed to stand for a time, a vigorous jet of water will wash off the solvent and all the grease and filth. If the dirt is thick and deeply embedded, work the solvent into it with a wire brush.

4. Finally wipe down the exterior of the engine

CHAPTER ONE

with a rag and only then, when it is quite clean should the dismantling process begin. As the engine is stripped, clean each part in a bath of paraffin or petrol.

5. Never immerse parts with oilways in paraffin, i.e. the crankshaft, but to clean wipe down carefully with a petrol dampened rag. Oilways can be cleaned out with pipe cleaners. If an air line is present all parts can be blown dry and the oilways blown through as an added precaution.

6. Re-use of old engine gaskets is a false economy and can give rise to oil and water leaks, if nothing worse. To avoid the possibility of trouble after the engine has been reassembled always use new gaskets throughout.

7. Do not throw the old gaskets away as it sometimes happens that an immediate replacement cannot be found and the old gasket is then very useful as a template. Hang up the old gaskets as they are removed on a suitable hook or nail.

8. To strip the engine it is best to work from the top down. The sump provides a firm base on which the engine can be supported in an upright position. When the stage where the sump must be removed is reached, the engine can be turned on its side and all other work carried out with it in this position.

9. Wherever possible, replace nuts, bolts and washers finger-tight from wherever they were removed. This helps avoid later loss and muddle. If they cannot be replaced then lay them out in such a fashion that it is clear from where they came.

10. If the engine was removed with the gearbox, separate them by undoing the nuts and bolts which hold the bellhousing to the engine endplate (photo).

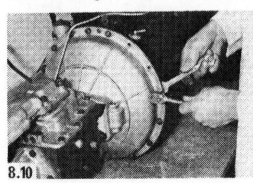

11. Also undo the two bolts holding the starter motor in place and lift off the motor (photo). Note and retain the distance piece and any shims which may be fitted.

12. Carefully pull the gearbox complete with bellhousing off the engine.

9. REMOVING ANCILLARY ENGINE COMPONENTS

1. Before basic engine dismantling begins the engine should be stripped of all its ancillary components. These items should also be removed if a factory exchange reconditioned unit is being purchased. The items comprise:-

Dynamo and dynamo brackets.
Water pump and thermostat housing.
Starter motor.
Distributor and sparking plugs.
Inlet and exhaust manifold and carburetters.
Fuel pump and fuel pipes.
Oil filter and dipstick.
Oil filler cap.
Clutch assembly.
Breather pipe and gauge (where fitted).
Auxiliary header tank (where fitted)

2. Without exception all these items can be removed with the engine in the car if it is merely an individual item which requires attention. (It is necessary to remove the gearbox if the clutch is to be renewed with the engine 'in situ').

3. Starting work on the left-hand side of the engine slacken off the dynamo retaining bolts (photo), and remove the unit and then the support brackets.

4. Take off the distributor and housing after undoing the two nuts and washers, (photo) which hold the bottom flange of the distributor housing to the cylinder block. Retain and note the shims between the housing and the block. Do not loosen the square nut on the clamp at the base of the distributor body or the timing will be lost. Undo the sparking plugs.

5. Note that the fuel pump is held in place by two studs. A nut fits on the stud on the left and a special screw over the stud on the right (photo).

6. Undo the nut and stud and lift off the fuel pump (photo).

7. Undo and remove the low oil pressure warning sender unit (photo).

8. Undo and remove the oil filter. The complete body just screws off anti-clockwise (photo).

9. Moving to the front of the engine undo the left-hand thermostat housing cover bolt to free the clip which carries the fuel and vacuum advance/retard lines.

10. Undo the nuts and washers which hold the inlet and exhaust manifolds to the cylinder head. The inner nuts are very difficult to get at, and are best loosened with a thin ring spanner (photo).

11. Lift off the inlet and exhaust manifolds together with the carburetter/s. If stiff tap the manifolds gently with a piece of wood. (Photo).

12. Undo the bolts which hold the water pump in place on the front face of the block. (Photo).

13. Undo the rear right cylinder head nut from the clip which holds the main heater pipe in place. (Photo).

14. The water pump is removed with the main heater pipe as one unit. (Photo).

15. Where a breather pipe is fitted note that it is a press fit in the block and should be carefully twisted and pulled out. (Photo).

16. Undo a quarter of a turn at a time the six bolts which hold the clutch pressure plate assembly to the flywheel. (Photo).

17. Lift off the pressure plate together with the loose friction plate. Check that all the items listed in paragraph 1 of this section have been removed. The engine is now stripped and ready for major dismantling to begin.

10. CYLINDER HEAD REMOVAL - ENGINE ON BENCH

1. With the engine out of the car and standing on its sump on the bench or on the floor remove the cylinder head as follows:-

2. Unscrew the two rocker cover nuts and lift the rocker cover and gasket away. (Photo).

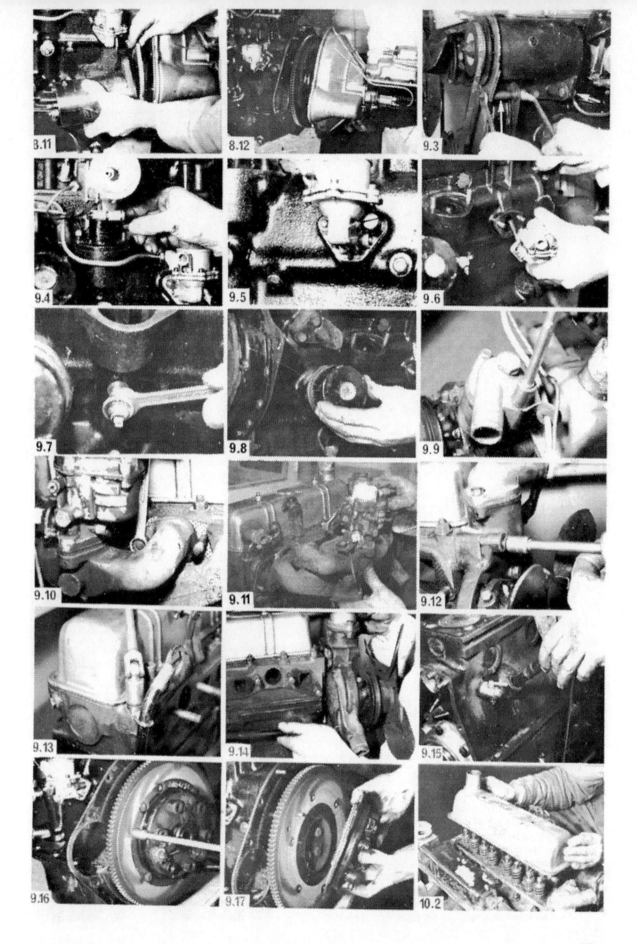

3. Unscrew the rocker pedestal nuts (four) and lift off the rocker assembly. (photo).

4. Undo the cylinder head nuts half a turn at a time in the reverse order shown in Fig. 1.3. When all the nuts are no longer under tension they may be screwed off the cylinder head one at a time.

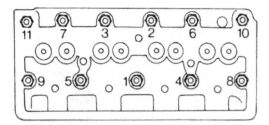

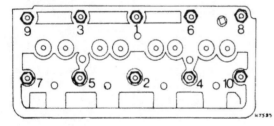

Fig. 1.3 Cylinder head nut tightening sequence
(alternative arrangements shown)

5. Remove the pushrods, keeping them in the relative order in which they were removed. The easiest way to do this is to push them through a sheet of thick paper or thin card in the correct sequence.

6. The cylinder head can now be removed by lifting upwards. If the head is jammed, try to rock it to break the seal. Under no circumstances try to prise it apart from the block with a screwdriver or cold chisel as damage may be done to the faces of the head or block. If the head will not readily free, turn the engine over by the flywheel as the compression in the cylinders will often break the cylinder head joint. If this fails to work, strike the head sharply with a plastic headed hammer, or with a wooden hammer, or with a metal hammer with an interposed piece of wood to cushion the blows. Under no circumstances hit the head directly with a metal hammer as this may cause the iron casting to fracture. Several sharp taps with the hammer at the same time pulling upwards should free the head. Lift the head off and place on one side. (Photo).

11. CYLINDER HEAD REMOVAL – ENGINE IN CAR

To remove the cylinder head with the engine still in the car the following additional procedure to that above must be followed. This procedure should be carried out before that listed in Section 10.

1. Disconnect the battery by removing the lead from the positive terminal. (Negative terminal on later cars.

2. Drain the water by turning the taps at the base of the radiator, and at the bottom left-hand corner of the cylinder block.

3. Loosen the clip at the thermostat housing end on the top water hose, and pull the hose from the thermostat housing pipe.

4. Slacken the dynamo securing bolts and move the dynamo away from the cylinder head (photo), at the same time remove the fan belt.

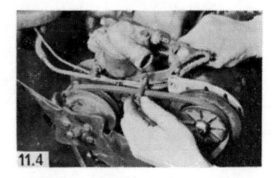

5. Disconnect the fuel line at the carburetter end, also the vacuum advance/retard pipe at the distributor and undo the three bolts holding the water pump and thermostat housing in place. (Photo).

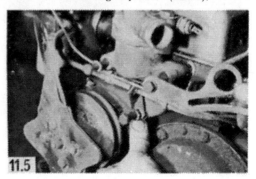

ENGINE

6. Remove the main heater pipe from the rear cylinder head stud (photo), and pull the water pump clear of the cylinder head.

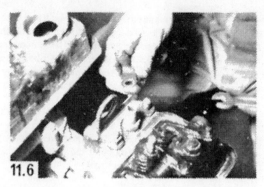

7. Disconnect the controls from the carburetter/s, and undo the three nuts holding the exhaust manifold to the exhaust down pipe. Leave the inlet and exhaust manifolds in place as they provide useful leverage when removing the cylinder head.
8. The procedure is now the same as for removing the cylinder head when on the bench. One tip worth noting is that should the cylinder head refuse to free easily, the battery can be reconnected up, and the engine turned over on the solenoid switch. Under no circumstances turn the ignition on and ensure the fuel inlet pipe is disconnected from the mechanical fuel pump.

12. VALVE REMOVAL

1. Two types of valve spring retainers are used on Triumph engines. The commonest type fitted is a special cap (see Fig. 1.5.) which has two interconnecting holes drilled in it, the larger of which slides over the valve stem. The double hole type is simply removed by pressing against the valve head with one hand, the other hand pressing the valve cap down at the same time pressing it across so the larger hole is directly around the valve stem so allowing the cap to come off.
2. The other type fitted is of normal split collet pattern. These are removed as follows:-
Compress each spring in turn with a valve spring compressor until the two halves of the collets can be removed. Release the compressor and remove the spring, shroud, and valve.
3. If, when the valve spring compressor is screwed down, the valve spring retaining cap refuses to free and expose the split collet, do not continue to screw down on the compressor as there is a likelihood of damaging it.
4. Gently tap the top of the tool directly over the cap with a light hammer. This will free the cap. To avoid the compressor jumping off the valve spring retaining cap when it is tapped, hold the compressor firmly in position with one hand. Drop each valve out through the combustion chamber.
5. It is essential that the valves are kept in their correct sequence unless they are so badly worn that they are to be renewed. If they are going to be kept and used again, place them in a sheet of card having eight holes numbered 1 to 8 corresponding with the relative positions the valves were in when fitted. Also keep the valve springs, washers, etc, in the correct order.

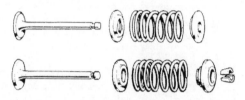

Fig. 1.5. The two types of valve spring retainers fitted to Triumph engines.

13. VALVE GUIDE REMOVAL

If it is wished to remove the valve guides they can be removed from the cylinder head in the following manner. Place the cylinder head with the gasket face on the bench and with a suitable hard steel punch drift the guides out of the cylinder head.

14. DISMANTLING THE ROCKER ASSEMBLY

1. To dismantle the rocker assembly, release the rocker shaft locating screw, remove the pins and caps, and spring washers from each end of the shaft and slide from the shaft the pedestals, rocker arms, and rocker spacing springs.
2. From the end of the shaft undo the plug which gives access to the inside of the rocker which can now be cleaned of sludge etc. Ensure the rocker arm lubricating holes are clear.

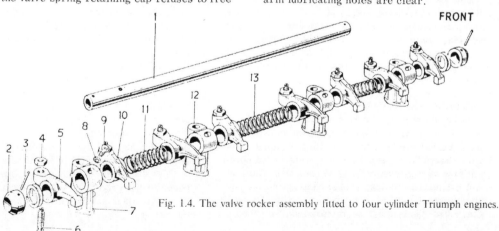

Fig. 1.4. The valve rocker assembly fitted to four cylinder Triumph engines.

CHAPTER ONE

15. TIMING COVER, GEARS & CHAIN REMOVAL

The timing cover, gears, and chain can be removed with the engine in the car providing the radiator and fan belt are first removed. The procedure for removing the timing cover, gears and chain is otherwise the same irrespective of whether the engine is in the car or on the bench, and is as follows:-

1. Bend back the locking tab of the crankshaft pulley locking washer under the crankshaft pulley retaining bolt, or starter dog bolt. (Photo).
2. This bolt is very large and it is unlikely that the average owner will possess a spanner large enough to undo it. To free the bolt place a metal drift in the position shown (photo) and hit the drift with a hammer until the bolt starts to turn.
3. The crankshaft pulley wheel may pull off quite easily. If not place two large screwdrivers behind the camshaft pulley wheel at 180° to each other, and carefully lever off the wheel. It is preferable to use a proper pulley extractor if this is available, but large screwdrivers or tyre levers are quite suitable, providing care is taken not to damage the pulley flange.
4. Remove the woodruff key from the crankshaft nose with a pair of pliers and note how the channel in the pulley is designed to fit over it. Place the woodruff key in a glass jam jar as it is a very small part and can easily become lost.
5. Unscrew the bolts holding the timing cover to the block. Note the special short screw adjacent to the crankshaft nose (photo).
6. Pull off the timing cover and gasket. The chain in the photo is very badly worn. On a less worn chain check for wear by measuring how much the chain can be depressed. More than $\frac{1}{2}$ in. means a new chain must be fitted on reassembly.
7. With the timing cover off, take off the oil thrower. NOTE that the concave side faces the gearwheel.
8. With a drift or screwdriver tap back the tabs on the lockwasher under the two camshaft gearwheel retaining bolts (photo), and undo the bolts.
9. To remove the camshaft and crankshaft timing wheels complete with chain, ease each wheel forward a little at a time levering behind each gearwheel in turn with two large screwdrivers at 180° to each other. If the gearwheels are locked solid then it will be necessary to use a proper gearwheel and pulley extractor, and if one is available this should be used anyway in preference to screwdrivers. With both gearwheels safely off, remove the woodruff key from the crankshaft with a pair of pliers and place them in the jam jar for safe keeping. Note the number of very thin packing washers behind the crankshaft gearwheel and remove them very carefully.

16. CAMSHAFT REMOVAL

The camshaft can be removed with the engine in place in the car, or with the engine on the bench. If the camshaft is to be removed with the engine in the car, the radiator, and fan belt must be removed after the cooling system has been drained. The inlet and exhaust manifolds, rocker gear, pushrods and tappets must also be removed. The timing cover, gears and chain, must be removed as described in Section 15. It is also necessary to remove the distributor drive gear as described in Section 17. With the drive gear out of the way, proceed in the following manner:-

1. First measure the camshaft endfloat with a feeler gauge placed between the keeper plate and the flange. If endfloat exceeds .008 in. it will be necessary to fit a new plate. Then remove the two bolts and spring washers which hold the camshaft locating plate to the block. The bolts are normally covered by the camshaft gearwheel.
2. Remove the plate (photo). The camshaft can now be withdrawn. Take great care to remove the camshaft gently, and in particular ensure that the cam peaks do not damage the camshaft bearings as the shaft is pulled forward.

17. DISTRIBUTOR DRIVE REMOVAL

1. To remove the distributor drive with the sump still in position first undo the two nuts which hold the distributor housing in place.
2. Lift off the distributor and distributor housing and then with a pair of long nosed pliers lift out the drive shaft (photo). As the shaft is removed turn it slightly to allow the shaft skew gears to disengage with the camshaft skew gear.

18. SUMP, PISTON, CONNECTING ROD & BIG END BEARING REMOVAL

1. The sump, pistons and connecting rods can be removed with the engine still in the car or with the engine on the bench. If in the car, proceed as for removing the cylinder head with the engine in the car, as described in Section 11. If on the bench proceed as for removing the cylinder head with the engine in this position, as described in Section 10. The pistons and connecting rods are drawn up out of the top of the cylinder bores.
2. Remove the bolts and washers holding the sump in position. Remove the sump and the sump gasket (photo).
3. Knock back with a cold chisel the locking tabs on the big end retaining bolts, and remove the bolts and locking tabs.
4. Remove the big end caps one at a time, taking care to keep them in the right order and the correct way round. Also ensure that the shell bearings are kept with their correct connecting rods and caps unless they are to be renewed. Normally, the numbers 1 to 4 are stamped on adjacent sides of the big end caps and connecting rods, indicating which cap fits on which rod and which way round the cap fits. If no numbers or lines can be found then with a sharp screwdriver or file scratch mating marks across the joint from the rod to the cap (photo). One line for connecting rod No.1, two for connecting rod No.2, and so on. This will ensure there is no confusion later as it is most important that the caps go back in the correct position on the connecting rods from which they were removed.
5. If the big end caps are difficult to remove they may be gently tapped with a soft hammer.
6. To remove the shell bearings, press the bearing opposite the groove in both the connecting rod, and the connecting rod caps and the bearings will slide out easily.

ENGINE

7. Withdraw the pistons and connecting rods upwards and ensure they are kept in the correct order for replacement in the same bore. Refit the connecting rod caps and bearings to the rods if the bearings do not require renewal to minimise the risk of getting the caps and rods muddled.

19. GUDGEON PIN - REMOVAL

1. To remove the gudgeon pin to free the piston from the connecting rod remove one of the circlips at either end of the pin with a pair of circlip pliers.
2. Press out the pin from the rod and piston with your finger.
3. If the pin shows reluctance to move, then on no account force it out, as this could damage the piston. Immerse the piston in a pan of boiling water for three minutes. On removal the expansion of the aluminium should allow the gudgeon pin to slide out easily.
4. Make sure the pins are kept with the same piston for ease of refitting.
5. Certain models use gudgeon pins which are an interference fit in the little end of the connecting rod. The tightness of fit is their sole means of retention. These pins must be pressed out and replaced using a special tool so as not to damage the connecting rods or the pistons. This is really a job best left to your local Triumph dealer or engineering works.

20. PISTON RING REMOVAL

1. To remove the piston rings, slide them carefully over the top of the piston, taking care not to scratch the aluminium alloy. Never slide them off the bottom of the piston skirt. It is very easy to break the iron piston rings if they are pulled off roughly so this operation should be done with extreme caution. It is helpful to make use of an old hacksaw blade, or better still, an old .020 in. feeler gauge.
2. Lift one end of the piston ring to be removed out of its groove and insert the end of the feeler gauge under it.
3. Turn the feeler gauge slowly round the piston and as the ring comes out of its groove apply slight upward pressure so that it rests on the land above. It can then be eased off the piston with the feeler gauge stopping it from slipping into any empty groove if it is any but the top piston ring that is being removed.

21. FLYWHEEL & ENGINE END PLATE REMOVAL

Having removed the clutch (see Chapter 5.5.) the flywheel and engine end plate can be removed. It is only possible for this operation to be carried out with the engine out of the car.

1. Bend back the locking tabs from the four bolts which hold the flywheel to the flywheel flange on the rear of the crankshaft.
2. Unscrew the bolts and remove them, complete with the locking plates (photo).
3. Lift the flywheel away from the crankshaft flange (photo). NOTE Some difficulty may be experienced in removing the bolts by the rotation of the crankshaft every time pressure is put on the spanner. To lock the crankshaft in position while the bolts are removed, use a screwdriver as a wedge between a

CHAPTER ONE

backplate stud and the ring gear as shown in the photo 21.2. Alternatively a wooden wedge can be inserted between the crankshaft and the side of the block inside the crankcase.

4. The engine endplate is held in position by a number of bolts and spring washers of varying size. Release the bolts noting where different sizes fit and place them together to ensure none of them become lost. Lift away the endplate from the block complete with the paper gasket. (Photo).

5. The front engine endplate is removed in identical fashion. (Photo)

22. CRANKSHAFT & MAIN BEARING REMOVAL

With the engine out of the car, remove the timing gears, sump, oil pump, and the big end bearings, pistons, flywheel and engine endplates as has already been described in Sections 15, 18 and 21. Removal of the crankshaft can only be attempted with the engine on the bench or floor. Take off the front sealing block and the packing pieces.

1. Undo by one turn the nuts which hold the three main bearing caps in place.

2. Unscrew the nuts and remove them together with the washers.

3. At the rear of the engine undo the seven bolts which hold the special oil retaining cover in place (photo) and remove the cover.

4. Remove the main bearing caps and the bottom half of each bearing shell, taking care to keep the bearing shells in the right caps.

5. When removing the rear bearing cap, NOTE the bottom semi-circular halves of the thrust washers, one half lying on either side of the main bearing. Lay them with the centre bearing along the correct side.

6. Slightly rotate the crankshaft to free the upper halves of the bearing shells and thrust washers which should now be extracted and placed over the correct bearing cap.

7. Remove the crankshaft by lifting it away from the crankcase (photo).

23. LUBRICATION & CRANKCASE VENTILATION SYSTEMS - DESCRIPTION

1. A forced feed system of lubrication is fitted with oil circulated round the engine from the sump below the block. The level of engine oil in the sump is indicated on the dipstick which is fitted on the right-hand side of the engine. It is marked to indicate the optimum level which is the maximum mark.

2. The level of the oil in the sump, ideally, should not be above or below this line. Oil is replenished via the filler cap on the rocker cover.

3. The eccentric rotor-type oil pump is bolted in the left-hand side of the crankcase and is driven by a short shaft from the skew gear on the camshaft which also drives the distributor shaft.

4. The pump is the non-draining variety to allow rapid pressure build-up when starting from cold.

5. Oil is drawn into the pump from the sump via the pick-up pipe. From the oil pump the lubricant passes through a non-adjustable relief valve (arrowed in photo) to the by-pass (early models only) or full flow filter. Filtered oil enters the main gallery which runs the length of the engine on the left-hand side. Drillings from the main gallery carry the oil to the crankshaft and camshaft journals.

6. The crankshaft is drilled so that oil under pressure reaches the crankpins from the crankshaft journals. The cylinder bores, pistons and gudgeon pins are all lubricated by splash and oil mist.

7. Oil is fed to the valve gear via the hollow rocker shaft at a reduced pressure by means of a scroll and two flats on the camshaft rear journal.

8. Drillings and grooves in the camshaft front journal lubricate the camshaft thrust plate, and the timing chain and gearwheels. Oil returns to the sump by gravity, the pushrods and cam followers being lubricated by oil returning via the pushrod drillings in the block.

9. Any one of three types of crankcase ventilation system may be fitted depending on the model and its year of manufacture. The three systems are known as 'Open Ventilation', 'Closed Ventilation', and 'Emission Control'.

10. 'Open ventilation' is very straightforward and is

ENGINE

only fitted to early models. It comprises an open angled tube fitted on the right-hand side of the engine which relieves crankcase pressure directly into the air.

11. 'Closed ventilation' is a slightly more sophisticated system with crankcase pressure being relieved by means of a rubber pipe from the rocker cover to the air cleaner. The hole for the open road tube is blocked over and the possibility of crankcase fumes entering the car is considerably reduced.

12. 'Emission control' is similar to 'closed ventilation' but more efficient and complicated. An emission control valve is positioned on top of the inlet manifold to which it is connected. It is also connected to a tube from the rocker cover. The control valve works by manifold depression so that when the depression is greatest (i.e. on the overrun) crankcase gas flow is restricted. A special oil filler cap is also used and this contains a non-return valve which ensures that crankcase and atmospheric pressures are kept in balance.

24. OIL FILTER - REMOVAL & REPLACEMENT

1. It is very easy to change the oil filter on all models.
2. Located on the left-hand side of the engine towards the front, unscrew the complete filter unit by grasping it firmly and turning it anticlockwise. (Photo).

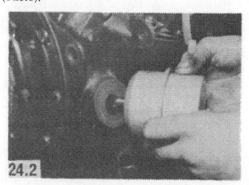

3. Throw the complete filter unit away, clean the mating faces on a new filter and the crankcase, and ensure the sealing ring on the new filter is undamaged.
4. Smear oil round the sealing ring (photo), and screw the new filter on tightly.

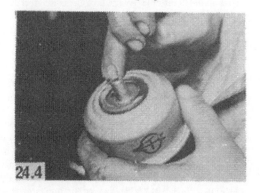

25. OIL PRESSURE RELIEF VALVE - REMOVAL & REPLACEMENT

1. To prevent excessive oil pressure - for example when the engine is cold - an oil pressure relief valve is built into the left-hand side of the engine immediately above the crankcase flange and in line vertically with the distributor.
2. The relief valve assembly is dismantled by undoing the large hexagon headed bolt which holds the relief valve piston and spring in place (photo).

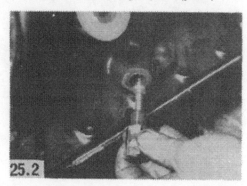

3. Always renew the spring at a major overhaul. To replace the assembly fit the valve piston into its orifice in the block, then the spring and then the bolt, ensuring that the sealing washer is in place on the latter.

26. OIL PUMP - REMOVAL & DISMANTLING

1. Undo the three bolts and spring washers which hold the pump to the block (photo).

2. Removal of the bolts also releases the end cover so the pump can be taken from the engine and the outer and inner rotors pulled off together with the pump shaft.

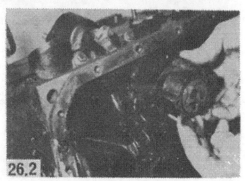

27. TIMING CHAIN TENSIONER - REMOVAL & REPLACEMENT

1. With time the spring blade timing chain tensioner will become worn and it should be renewed at the same time as the timing chain. Wear can be clearly seen as two grooves on the face of the tensioner where it presses against the chain.(Photo).

2. To remove the tensioner bend it back and then pull out from its securing pins (Photo).

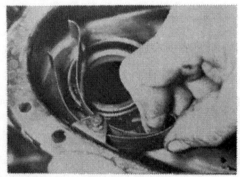

3. On replacement fit the open end of the tensioner over the pin and press the blade into place with the aid of a screwdriver until it snaps into place (photo).

28. EXAMINATION & RENOVATION - GENERAL

With the engine stripped down and all parts thoroughly cleaned, it is now time to examine everything for wear. The following items should be checked and where necessary renewed or renovated as described in the following sections.

29. CRANKSHAFT EXAMINATION & RENOVATION

Examine the crankpin and main journal surfaces for signs of scoring or scratches. Check the ovality of the crankpins at different positions with a micrometer. If more than 0.001 in. out of round, the crankpins will have to be reground. It will also have to be reground if there are any scores or scratches present. Also check the journals in the same fashion. On highly tuned engines the centre main bearing has been known to break up. This is not always immediately apparent, but slight vibration in an otherwise normally smooth engine and a very slight drop in oil pressure under normal conditions are clues. If the centre main bearing is suspected of failure it should be immediately investigated by dropping the sump and removing the centre main bearing cap. Failure to do this will result in a badly scored centre main journal. If it is necessary to regrind the crankshaft and fit new bearings your local Triumph garage or engineering works will be able to decide how much metal to grind off and the correct undersize shells to fit.

30. BIG END & MAIN BEARINGS - EXAMINATION & RENOVATION

Big end bearing failure is accompanied by a noisy knocking from the crankcase, and a slight drop in oil pressure. Main bearing failure is accompanied by vibration which can be quite severe as the engine speed rises and falls and a drop in oil pressure.

Bearings which have not broken up, but are badly worn will give rise to low oil pressure and some vibration. Inspect the big ends, main bearings, and thrust washers for signs of general wear, scoring, pitting and scratches. The bearings should be matt grey in colour. With lead-indium bearings should a trace of copper colour be noticed the bearings are badly worn as the lead bearing material has worn away to expose the indium underlay. Renew the bearings if they are in this condition or if there is any sign of scoring or pitting.

The undersizes available are designed to correspond with the regrind sizes, i.e. - .010 in. bearings are correct for a crankshaft reground - .010 in. undersize. The bearings are in fact, slightly more than the stated undersize as running clearances have been allowed for during their manufacture.

Very long engine life can be achieved by changing big end bearings at intervals of 30,000 miles and main bearings at intervals of 50,000 miles, irrespective of bearing wear. Normally, crankshaft wear is infinitesimal and a change of bearings will ensure mileages of between 100,000 to 120,000 miles before crankshaft regrinding becomes necessary. Crankshafts normally have to be reground because of scoring due to bearing failure.

31. CYLINDER BORES - EXAMINATION & RENOVATION

1. The cylinder bores must be examined for taper, ovality, scoring and scratches. Start by carefully examining the top of the cylinder bores. If they are at all worn a very slight ridge will be found on the thrust side. This marks the top of the piston ring travel. The owner will have a good indication of the bore wear prior to dismantling the engine, or removing the cylinder head. Excessive oil consumption accompanied by blue smoke from the exhaust is a sure sign of worn cylinder bores and piston rings.
2. Measure the bore diameter just under the ridge

with a micrometer and compare it with the diameter at the bottom of the bore, which is not subject to wear. If the difference between the two measurements is more than .006 in. then it will be necessary to fit special pistons and rings or to have the cylinders rebored and fit oversize pistons. If no micrometer is available remove the rings from a piston and place the piston in each bore in turn about 3/4 in. below the top of the bore. If an 0.010 feeler gauge can be slid between the piston and the cylinder wall on the thrust side of the bore then remedial action must be taken. Oversize pistons are available in the following sizes:-

+ .010 in. (254 mm.), + .020 in. (.508 mm.), + .030 in. (.762 mm.).

3. These are accurately machined to just below these measurements so as to provide correct running clearances in bores bored out to the exact oversize dimensions.

4. If the bores are slightly worn but not so badly worn as to justify reboring them, then special oil control rings and pistons can be fitted which will restore compression and stop the engine burning oil. Several different types are available and the manufacturers instructions concerning their fitting must be followed closely.

5. If the block is to be sent away for reboring it is essential to remove the cylinder head studs. Lock two nuts together on a stud (photo) and then wind the stud out by turning the bottom nut anti-clockwise.

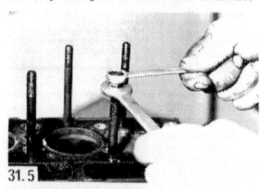

6. If new pistons are being fitted and the bores have not been reground, it is essential to slightly roughen the hard glaze on the sides of the bores with fine glass paper (photo) so the new piston rings will have a chance to bed in properly.

32. PISTONS & PISTON RINGS - EXAMINATION & RENOVATION

If the old pistons are to be refitted, carefully remove the piston rings and then thoroughly clean them. Take particular care to clean out the piston ring grooves. At the same time do not scratch the aluminium in any way. If new rings are to be fitted to the old pistons then the top ring should be stepped so as to clear the ridge left above the previous top ring. If a normal but oversize new ring is fitted, it will hit the ridge and break, because the new ring will not have worn in the same way as the old, which will have worn in unison with the ridge.

Before fitting the rings on the pistons each should be inserted approximately 3 in. down the cylinder bore and the gap measured with a feeler gauge. This should be between .015 in. and .038 in. It is essential that the gap should be measured at the bottom of the ring travel, as if it is measured at the top of a worn bore and gives a perfect fit, it could easily seize at the bottom. If the ring gap is too small rub down the ends of the ring with a very fine file until the gap, when fitted, is correct. To keep the rings square in the bore for measurement line each up in turn by inserting an old piston in the bore upside down, and use the piston to push the ring down about 3 in. Remove the piston and measure the piston ring gap.

When fitting new pistons and rings to a rebored engine the piston ring gap can be measured at the top of the bore as the bore will not now taper. It is not necessary to measure the side clearance in the piston ring grooves with the rings fitted as the groove dimensions are accurately machined during manufacture. When fitting new oil control rings to old pistons it may be necessary to have the grooves widened by machining to accept the new wider rings. In this instance the manufacturers representative will make this quite clear and will supply the address to which the pistons must be sent for machining.

33. CAMSHAFT & CAMSHAFT BEARINGS - EXAMINATION & RENOVATION

On the majority of engines the camshaft runs direct in the block and wear of the journals and bearings is negligable. On certain of the latest models pre-formed camshaft bearings are fitted and these can be replaced. This is an operation for your Triumph dealer or the local engineering works as it demands the use of specialised equipment. The bearings are removed with a special drift after which new bearings are pressed in, care being taken to ensure the oil holes in the bearings line up with those in the block. On no account can the bearings be reamed in position.

The camshaft itself should show no signs of wear, but, if very slight scoring on the cams is noticed, the score marks can be removed by very gentle rubbing down with a very fine emery cloth. The greatest care should be taken to keep the cam profiles smooth.

34. VALVES & VALVE SEATS - EXAMINATION & RENOVATION

1. Examine the heads of the valves for pitting and burning, especially the heads of the exhaust valves.

The valve seatings should be examined at the same time. If the pitting on valve and seat is very slight the marks can be removed by grinding the seats and valves together with coarse, and then fine, valve grinding paste. Where bad pitting has occured to the valve seats it will be necessary to recut them and fit new valves. If the valve seats are so worn that they cannot be recut, then it will be necessary to fit new valve seat inserts. These latter two jobs should be entrusted to the local Triumph agent or engineering works. In practice it is very seldom that the seats are so badly worn that they require renewal. Normally, it is the exhaust valve that is too badly worn for replacement, and the owner can easily purchase a new set of valves and match them to the seats by valve grinding.

2. Valve grinding is carried out as follows:-

Smear a trace of coarse carborundum paste on the seat face and apply a suction grinder tool to the valve head. With a semi-rotary motion, grind the valve head to its seat, lifting the valve occasionally to redistribute the grinding paste. (Photo). When

a dull matt even surface finish is produced on both the valve seat and the valve, then wipe off the paste and repeat the process with fine carborundum paste, lifting and turning the valve to redistribute the paste as before. A light spring placed under the valve head will greatly ease this operation. When a smooth unbroken ring of light grey matt finish is produced, on both valve and valve seat faces, the grinding operation is completed.

3. Scrape away all carbon from the valve head and the valve stem. Carefully clean away every trace of grinding compound, taking great care to leave none in the ports or in the valve guides. Clean the valves and valve seats with a paraffin soaked rag then with a clean rag, and finally, if an air line is available, blow the valves, valve guides and valve ports clean.

35. TIMING GEARS & CHAIN - EXAMINATION & RENOVATION

Examine the teeth on both the crankshaft gearwheel and the camshaft gearwheel for wear. Each tooth forms an inverted 'V' with the gearwheel periphery, and if worn the side of each tooth under tension will be slightly concave in shape when compared with the other side of the tooth, i.e. one side of the inverted 'V' will be concave when compared with the other. If any sign of wear is present the gearwheels must be renewed.

Examine the links of the chain for side slackness and renew the chain if any slackness is noticeable when compared with a new chain. It is a sensible precaution to renew the chain at about 30,000 miles and at a lesser mileage if the engine is stripped down for a major overhaul. The actual rollers on a very badly worn chain may be slightly grooved.

36. TIMING CHAIN TENSIONER - EXAMINATION & RENOVATION

1. If the timing chain is badly worn it is more than likely that the tensioner will be too.
2. Examine the side of the tensioner which bears against the chain and renew it if it is grooved or ridged. See Section 27 for details.

37. ROCKERS & ROCKER SHAFT - EXAMINATION & RENOVATION

Remove the threaded plug with a screwdriver from the end of the rocker shaft and thoroughly clean out the shaft. As it acts as the oil passage for the valve gear also ensure the oil holes in it are quite clear after having cleaned them out. Check the shaft for straightness by rolling it on the bench. It is most unlikely that it will deviate from normal, but, if it does, then a judicious attempt must be made to straighten it. If this is not successful purchase a new shaft. The surface of the shaft should be free from any worn ridges caused by the rocker arms. If any wear is present, renew the shaft. Wear is only likely to have occured if the rocker shaft oil holes have become blocked.

Check the rocker arms for wear of the rocker bushes, for wear at the rocker arm face which bears on the valve stem, and for wear of the adjusting ball ended screws. Wear in the rocker arm bush can be checked by gripping the rocker arm tip and holding the rocker arm in place on the shaft, noting if there is any lateral rocker arm shake. If shake is present, and the arm is very loose on the shaft, a new bush or rocker arm must be fitted.

Check the tip of the rocker arm where it bears on the valve head for cracking or serious wear on the case hardening. If none is present reuse the rocker arm. Check the lower half of the ball on the end of the rocker arm adjusting screw. On high performance engines wear on the ball and top of the pushrod is easily noted by the unworn 'pip' which fits in the small central oil hole on the ball. The larger this 'pip' the more wear has taken place to both the ball and the pushrod. Check the pushrods for straightness by rolling them on the bench. Renew any that are bent.

38. TAPPETS - EXAMINATION & RENOVATION

Examine the bearing surface of the tappets which lie on the camshaft. Any indentation in this surface or any cracks indicate serious wear and the tappets should be renewed. Thoroughly clean them out, removing all traces of sludge. It is most unlikely that the sides of the tappets will prove worn, but, if they are a very loose fit in their bores and can readily be rocked, they should be exchanged for new units. It is very unusual to find any wear in the tappets, and any wear present is likely to occur only at very high mileages.

ENGINE

39. FLYWHEEL STARTER RING - EXAMINATION & RENOVATION

If the teeth on the flywheel starter ring are badly worn, or if some are missing, then it will be necessary to remove the ring. This is achieved by splitting the ring with a cold chisel. The greatest care should be taken not to damage the flywheel during this process.

To fit a new ring heat it gently and evenly with an oxyacetylene flame until a temperature of approximately $350^{o}C$ is reached. This is indicated by a light metallic blue surface colour. With the ring at this temperature, fit it to the flywheel with the front of the teeth facing the flywheel register. The ring should be tapped gently down onto its register and left to cool naturally when the shrinkage of the metal on cooling will ensure that it is a secure and permanent fit. Great care must be taken not to overheat the ring, as if this happens the temper of the ring will be lost.

40. OIL PUMP - EXAMINATION & RENOVATION

Thoroughly clean all the component parts in petrol and then check the rotor endfloat and lobe clearances in the following manner:-

Position the rotors in the pump and place the straight edge of a steel ruler across the joint face of the pump. Measure the gap between the bottom of the straight edge and the top of the rotors with a feeler gauge as at 'A' in Fig.1.6. If the measurement exceeds .005 in. (.127 mm.) then check the lobe clearances as described in the following paragraphs. If the lobe clearances are correct then lap the joint face on a sheet of plate glass.

Measure with a feeler gauge the gap between the inner and outer rotors. It should not be more than 0.010 in.

Then measure the gap between the outer rotor and the side of the pump body which should not exceed 0.008 in. It is essential to renew the pump if the measurements are outside these figures. It can be safely assumed that at any major reconditioning the pump will need renewal.

41. CYLINDER HEAD - DECARBONISATION

This can be carried out with the engine either in or out of the car. With the cylinder head off carefully remove with a wire brush and blunt scraper all traces of carbon deposits from the combustion spaces and the ports. The valve head stems and valve guides should also be freed from any carbon deposits. Wash the combustion spaces and ports down with petrol and scrape the cylinder head surface free of any foreign matter with the side of a steel rule, or a similar article.

Clean the pistons and top of the cylinder bores. If the pistons are still in the block then it is essential that great care is taken to ensure that no carbon gets into the cylinder bores as this could scratch the cylinder walls or cause damage to the piston and rings. To ensure this does not happen, first turn the crankshaft so that two of the pistons are at the top of their bores. Stuff rag into the other two bores or seal them off with paper and masking tape. The waterways should also be covered with small pieces of masking tape to prevent particles of carbon entering the cooling system and damaging the water pump.

There are two schools of thought as to how much carbon should be removed from the piston crown. One school recommends that a ring of carbon should be left round the edge of the piston and on the cylinder bore wall as an aid to low oil consumption. Although this is probably true for early engines with worn bores, on later engines the thought of the second school can be applied; which is that for effective decarbonisation all traces of carbon should be removed.

If all traces of carbon are to be removed, press a little grease into the gap between the cylinder walls and the two pistons which are to be worked on. With a blunt scraper carefully scrape away the carbon from the piston crown, taking great care not to scratch the aluminium. Also scrape away the carbon from the surrounding lip of the cylinder wall. When all carbon has been removed, scrape away the grease which will now be contaminated with

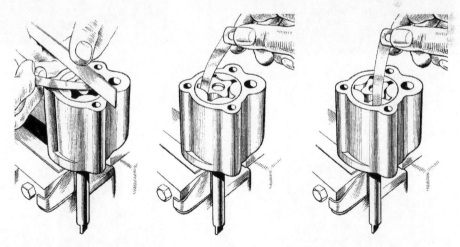

Fig. 1.6. Oil pump checking rotor endfloat and lobe clearances.

carbon particles, taking care not to press any into the bores. To assist prevention of carbon build-up the piston crown can be polished with a metal polish such as Brasso. Remove the rags or masking tape from the other two cylinders and turn the crankshaft so that the two pistons which were at the bottom are now at the top. Place rag or masking tape in the cylinders which have been decarbonised and proceed as just described.

If a ring of carbon is going to be left round the piston then this can be helped by inserting an old piston ring into the top of the bore to rest on the piston and ensure that carbon is not accidentally removed. Check that there are no particles of carbon in the cylinder bores. Decarbonising is now complete.

42. VALVE GUIDES - EXAMINATION & RENOVATION

Examine the valve guides internally for wear. If the valves are a very loose fit in the guides and there is the slightest suspicion of lateral rocking using a new valve, then new guides will have to be fitted. If the valve guides have been removed compare them internally by visual inspection with a new guide as well as testing them for rocking with a new valve

43. SUMP - EXAMINATION & RENOVATION

1. It is essential to thoroughly wash out the sump with petrol and this can only be done properly with the gauze removed.
2. With a screwdriver and a pair of pliers carefully pull back the tags which hold the gauze in place (photo).

3. The gauze can then be lifted out (photo) and the inside cleaned out properly. Scrape all traces of the old sump gasket from the flange.

44. ENGINE REASSEMBLY - GENERAL

1. To ensure maximum life with minimum trouble from a rebuilt engine, not only must everything be correctly assembled, but all the parts must be spotlessly clean, all the oilways must be clear, locking washers and spring washers must always be fitted where indicated and all bearing and other working surfaces must be thoroughly lubricated during assembly. Before assembly begins renew any bolts or studs the threads of which are in any way damaged, and whenever possible use new spring washers.
2. Check the core plugs for signs of weeping and always renew the plug at the front of the engine as it is normally covered by the engine endplate.
3. Drive a punch through the centre of the core plug. (Photo).

4. Using the punch as a lever lift out the old core plug. (Photo).

5. Thoroughly clean the core plug orifice and using a thin headed hammer as an expander firmly tap a new core plug in place, convex side facing out. (Photo)

ENGINE

6. Apart from your normal tools, a supply of clean rag, an oil can filled with engine oil (an empty plastic detergent bottle thoroughly cleaned and washed out, will invariably do just as well), a new supply of assorted spring washers, a set of new gaskets, and preferably a torque spanner, should be collected together.

45. CRANKSHAFT REPLACEMENT

Ensure that the crankcase is thoroughly clean and that all oilways are clear. A thin-twist drill or a pipe cleaner is useful for cleaning them out. If possible, blow them out with compressed air.

Treat the crankshaft in the same fashion, and then inject engine oil into the crankshaft oilways.

Commence work on rebuilding the engine by replacing the crankshaft and main bearings:-

1. If the old main bearing shells are to be replaced, (a false economy unless they are virtually as new), fit the three upper halves of the main bearing shells to their location in the crankcase, after wiping the locations clean (photo).
2. NOTE that at the back of each bearing is a tab which engages in locating grooves in either the crankcase or the main bearing cap housings.
3. If new bearings are being fitted, carefully clean away all traces of the protective grease with which they are coated.
4. With the three upper bearing shells securely in place, wipe the lower bearing cap housings and fit the three lower shell bearings to their caps ensuring that the right shell goes into the right cap if the old bearings are being refitted.
5. Wipe the recesses either side of the rear main bearing which locate the thrust washers.
6. Generously lubricate the crankshaft journals and the upper and lower main bearing shells and carefully lower the crankshaft into place. (Photo).
7. Fit the upper halves of the thrust washers into their grooves either side of the rear main bearing (photo), rotating the crankshaft in the direction towards the main bearing tabs (so that the main bearing shells do not slide out). At the same time feed the thrust washers into their locations with their oil grooves outwards away from the bearing.
8. Fit the main bearing caps in position ensuring they locate properly. The mating surfaces must be spotlessly clean or the caps will not seat correctly (photo). As the bearing caps were assembled to the cylinder block and then line bored during manufacture, it is essential that they are returned to the same positions from which they were removed.
9. Refit the main bearing cap bolts and locking tabs (if fitted) and tighten the bolts to a torque of 55 lb. ft. (Photo).
10. Test the crankshaft for freedom of rotation, should it be very stiff to turn or posses high spots a most careful inspection must be made, preferably by a qualified mechanic with a micrometer to get to the root of the trouble. It is very seldom that any trouble of this nature will be experienced when fitting the crankshaft.
11. Check the crankshaft endfloat with a feeler gauge measuring the longitudinal movement between the crankshaft and a bearing cap. Endfloat should be between 0.004 in. and 0.008 in. If endfloat is excessive oversize thrust washers can be fitted.
12. Next fit the sealing block over the front main bearing cap. Smear the ends of the block with jointing compound (photo), and fit the block in place. Fit the securing screws but do not tighten fully. Fit new wedge seals at each end and line up the front face of the block with the front of the cylinder block. Tighten the screws fully and cut the wedge seals flush with the crankcase flange.

46. PISTON & CONNECTING ROD REASSEMBLY

1. If the same pistons are being used, then they must be mated to the same connecting rod with the same gudgeon pin. If new pistons are being fitted it does not matter which connecting rod they are used with, but, the gudgeon pins should be fitted on the basis of selective assembly.
2. If interference fit gudgeon pins are used on your engine (i.e. the pins are held firmly in place by tightness of fit in the little end), then the pistons

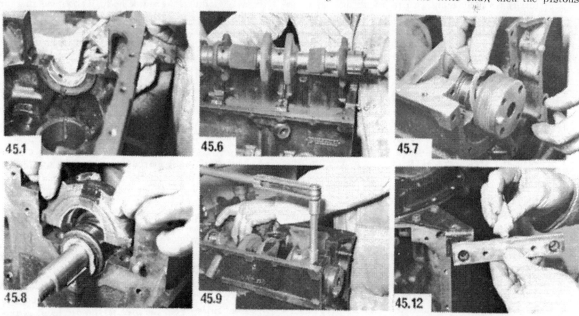

CHAPTER ONE

and connecting rods must be assembled at your local Triumph agent who will have the special tool necessary to draw the pin in, in conjunction with a torque wrench.

3. Because aluminium alloy, when hot, expands more than steel, the gudgeon pin may be a very tight fit in the piston when they are cold. To avoid any damage to the piston it is best to heat it in boiling water when the pin will slide in easily.

4. Lay the correct piston adjacent to each connecting rod and remember that the same rod and piston must go back into the same bore. If new pistons are being used it is only necessary to ensure that the right connecting rod is placed in each bore.

5. Fit a gudgeon pin circlip in position at one end of the gudgeon pin hole in the piston.

6. Locate the connecting rod in the piston with the marking 'FRONT' on the piston crown towards the front of the engine, i.e. the timing cover end, and the connecting rod cap towards the camshaft side of the engine.

7. Slide the gudgeon pin in through the hole in the piston and through the connecting rod little end until it rests against the previously fitted circlip (photo). NOTE that the pin should be a push fit.

8. Fit the second circlip in position (photo). Repeat this procedure for all four pistons and connecting rods.

9. Where special oil control pistons are being fitted should the position of the top ring be the same as the position of the top ring on the old piston ensure that a groove has been machined on the top of the new ring so no fouling occurs between the unworn portion at the top of the bore and the piston ring when the latter is at the top of its stroke (photo).

47. PISTON RING REPLACEMENT

1. Check that the piston ring grooves and oilways are thoroughly clean and unblocked. Piston rings must always be fitted over the head of the piston and never from the bottom.

2. The easiest method to use when fitting rings is to wrap a .020 feeler gauge round the top of the piston and place the rings one at a time, starting with the bottom oil control ring, over the feeler gauge.

3. The feeler gauge, complete with ring, can then be slid down the piston over the other piston ring grooves until the correct groove is reached. The piston ring is then slid gently off the feeler gauge into the groove.

4. An alternative method is to fit the rings by holding them slightly open with the thumbs and both of your index fingers. This method requires a steady hand and great care as it is easy to open the ring too much and break it.

48. PISTON REPLACEMENT

The pistons, complete with connecting rods, can be fitted to the cylinder bores in the following sequence:-

1. With a wad of clean rag wipe the cylinder bores clean.

2. The pistons, complete with connecting rods, are fitted to their bores from the top of the block (photo).

3. As each piston is inserted into its bore ensure that it is the correct piston/connecting rod assembly for that particular bore and that the connecting rod is the right way round, and that the front of the piston is towards the front of the bore, i.e. towards the front of the engine.

4. The piston will only slide into the bore as far as the oil control ring. It is then necessary to compress the piston rings into a clamp (photo) and to gently tap the piston into the cylinder bore with a wooden or plastic hammer. If a proper piston ring clamp is not available then a suitable jubilee clip does the job very well.

49. CONNECTING ROD TO CRANKSHAFT REASSEMBLY

1. Wipe the connecting rod half of the big end bearing cap and the underside of the shell bearing clean, and fit the shell bearing in position with its locating tongue engaged with the corresponding rod (photo).

2. If the old bearings are nearly new and are being refitted then ensure they are replaced in their correct locations on the correct rods.

3. Generously lubricate the crankpin journals with engine oil (photo), and turn the crankshaft so that the crankpin is in the most advantageous position for the connecting rod to be drawn onto it.

4. Wipe the connecting rod bearing cap and back of the shell bearing clean and fit the shell bearing in position ensuring that the locating tongue at the back of the bearing engages with the locating groove in connecting rod cap

5. Generously lubricate the shell bearing and offer up the connecting rod bearing cap to the connecting rod. (See photo).

6. Fit the connecting rod bolts with the one-piece locking tab under them and tighten the bolts with a torque spanner to 40 lb/ft. (Photo).

7. With a hammer or pair of pliers knock up the locking tabs against the bolt head. (Photo).

8. When all the connecting rods have been fitted, rotate the crankshaft to check that everything is free, and that there are no high spots causing binding. The bottom half is now nearly built up (photo).

50. FRONT END PLATE - REASSEMBLY

1. Fit a new gasket in place over the front of the cylinder block (photo).

2. Lower the front end plate into place noting the hole for the dowel (arrowed) and then fit the securing bolt immediately above the crankshaft nose (photo).

51. CAMSHAFT REPLACEMENT

1. Wipe the camshaft bearing journals clean and lubricate them generously with engine oil.

2. Insert the camshaft into the crankcase gently (photo) taking care not to damage the camshaft bearings with the cams.

3. Replace the camshaft locating plate and tighten down the two retaining bolts and washers. (Photo).

52. TIMING GEARS - CHAIN TENSIONER - COVER REPLACEMENT

1. Place the gearwheels in position (photo) without the timing chain and place the straight edge of a steel ruler from the side of the camshaft gearteeth

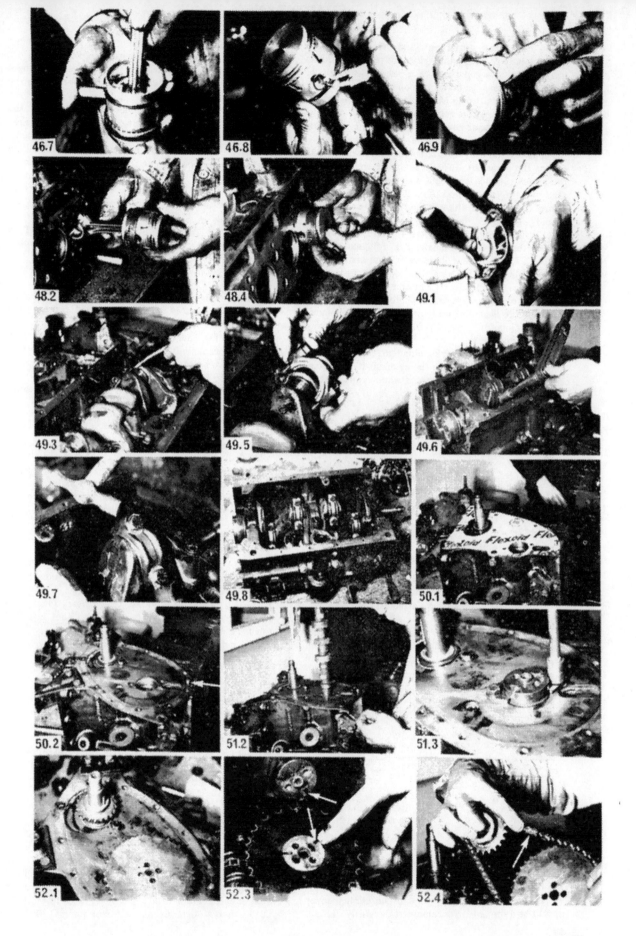

to the crankshaft gearwheel, and measure the gap (if any) between the steel rule and the crankshaft gearwheel. If a gap exists a suitable number of packing washers must be placed on the crankshaft nose to bring the crankshaft gearwheel onto the same plane as the camshaft gearwheel.

2. Fit the woodruff key to the slot in the crankshaft nose.

3. It is all too easy to fit the sprocket wheel 180° out on the camshaft. The best way of ensuring that the wheel is fitted the right way round is to ensure the two different slot marks on the back of the wheel correspond with the slots on the front of the camshaft (photo).

4. Lay the camshaft gearwheels on a clean surface so that the two timing marks are adjacent to each other. Slip the timing chain over them and pull the gearwheels back into mesh with the chain so that the timing marks, although further apart, are still adjacent to each other, (as in photo 52.6). A special point to note is that should the chain have a removable link (photo), always position it so the spring clip faces forwards.

5. With the timing marks adjacent to each other hold the gearwheels above the crankshaft and camshaft (photo). Turn the camshaft and crankshaft so

that the woodruff key will enter the slot in the crankshaft gearwheel, and the camshaft gearwheel is in the correct position relative to the camshaft, (see paragraph 3).

6. Fit the timing chain and gearwheel assembly onto the camshaft and crankshaft, keeping the timing marks adjacent, (photo). Fit a new double tab

washer in place on the camshaft gearwheel and fit the two retaining bolts.

7. Lever up the tabs on the lockwasher (photo).

8. The oil seal in the front of the timing cover should be renewed. To remove it carefully drive it out with a screwdriver taking care not to damage the timing cover in the process (photo).

9. Evenly press a new seal into the cover using a vice (photo) ensuring that the seal lip is towards the crankshaft sprocket wheel.

10. Fit the oil thrower in place on the nose of the crankshaft (photo) making sure that the dished periphery is towards the cover (if dished type fitted).

11. Lubricate the front cover oil seal, fit a new gasket in place on the end plate, and fit the cover at an angle (photo), so as to catch the spring tensioner against the side of the chain. Swing the cover into its correct position and insert one or two bolts finger tight.

12. Note that the short screw headed bolt MUST be fitted to the hole indicated by the arrow in the photograph.

13. Now tighten down all the bolts and screws evenly (photo).

14. Then fit the crankshaft nose pulley wheel (photo).

15. The next step is to replace the pulley wheel nut or starter dog (photo) using a new lockwasher.

16. Tighten the nut or starter dog (photo A) and prevent the crankshaft from moving by temporarily refitting two bolts and holding a screwdriver between them (photo B).

17. Finally knock down one of the lockwasher tabs, (photo).

53. OIL PUMP REPLACEMENT

1. Fit the pump and drive shaft to the crankcase, (photo).

2. Prime the pump to preclude any possibility of oil starvation when the engine starts (photo).

3. Refit the cover to the pump and tighten down the three securing bolts and washers. (Photo).

54. CRANKSHAFT REAR SEAL, HOUSING, ENDPLATE & FLYWHEEL REPLACEMENT

1. A scroll type crankshaft rear oil seal was used on early models. Later models are fitted with a lip type seal.

2. To fit the scroll type seal coat a new gasket with jointing compound, position it on the seal housing (photo) and fit the housing to the crankcase, doing up the retaining bolts and spring washers finger tight.

3. Check with a feeler gauge that a gap of 0.003 in. (0.076 mm.), (aluminium housings only) exists all round the crankshaft journal, tapping the housing with a soft headed hammer until the seal is centralised (photo). Some later models make use of a cast iron housing. This is fitted in just the same way but the clearance should be 0.002 in. (0.508 mm.)

4. To fit the lip type seal first coat both sides of a new gasket with jointing compound and position the gasket on the crankcase joint face.

5. Press a new seal into the crankshaft housing with the lip of the seal facing the crankshaft. Oil the seal and carefully fit the housing making sure the lip of the seal is not turned over. Replace the housing bolts finger tight, turn the crankshaft over several times to centralise the seal, and tighten the bolts down firmly. Irrespective of what type of seal is used now fit the input shaft bush to the hole in the centre of the crankshaft rear journal (photo).

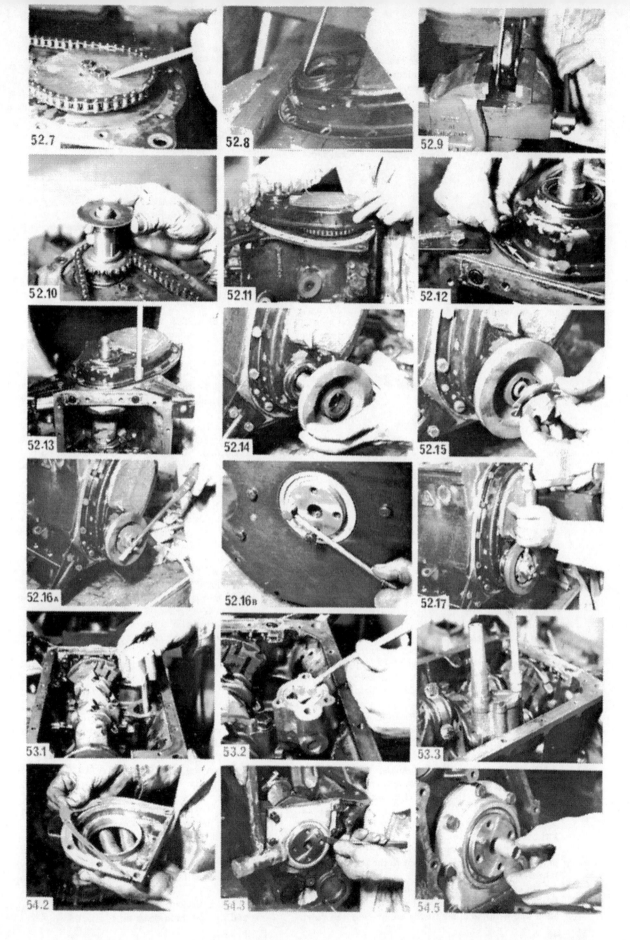

CHAPTER ONE

6. No gasket is fitted between the end plate and the block. Fit the end plate in place and tighten down the bolts and washers. (Photo).

54.6

7. Make certain that the flange on the crankshaft, and the face of the flywheel are perfectly clean and offer up the flywheel to the end of the crankshaft. Ensure that the dowel enters into the special hole in the flywheel. Fit new tab washers, tighten down the four retaining bolts (photo) and turn up the lock tags.

54.7

8. Smear the crankshaft spigot bush with a small quantity of zinc oxide grease. (Photo).

55. SUMP REPLACEMENT

1. After the sump has been thoroughly cleaned, scrape all traces of the old sump gasket from the sump and crankcase flanges, fit a new gasket in placed, and then refit the sump. (Photo).
2. Insert and tighten down the sump bolts and washers remembering to fit the bracket to the third bolt down on the right-hand side. (Photo).

56. VALVE & VALVE SPRING REASSEMBLY

To refit the valves and valve springs to the cylinder head, proceed as follows:-
1. Rest the cylinder head on its side.
2. Fit each valve and valve spring in turn, wiping down and lubricating each valve stem as it is inserted into the same valve guide from which it was removed. (Photo).
3. Build up each valve assembly by first fitting the lower collar. (Photo).
4. Then fit the valve spring so that the qlosely coiled portion of the spring is adjacent to the cylinder head (photo).
5. On engines which use the double hole spring retaining collar press the valve in firmly with one hand, and with the other fit the collar over the valve stem by means of the offset hole (photo).
6. Press down hard on the collar to compress the spring and as soon as the collar is in line with the groove in the valve stem push the collar across into the smaller hole so the spring is securely retained. (Photo).
7. On engines which use split collets to retain the upper retaining collar, move the cylinder head towards the edge of the work bench if it is facing downwards and slide it partially over the edge of the bench so as to fit the bottom half of a valve spring compressor to the valve head. Slide the spring and upper collar over the valve stem.
8. With the base of the valve compressor on the valve head, compress the valve spring until the cotters can be slipped into place in the cotter grooves. Gently release the compressor.
9. Repeat this procedure until all eight valves and valve springs are fitted.

57. ROCKER SHAFT & TAPPET REASSEMBLY

1. Fit an end cap and pin to one end of the shaft and then slide on the springs, rockers, distance springs, and rocker pedestals in their correct order as shown in Fig. 1.4.
2. Make sure that the Phillips screw on the rear rocker pedestal engages properly with the rocker shaft.
3. When all is correctly assembled fit the remaining end cap and oil the components thoroughly.
4. Generously lubricate the tappets internally and externally and insert them in the bores from which they were removed. (Photo).

58. CYLINDER HEAD REPLACEMENT

1. Thoroughly clean the cylinder block top face and then refit the cylinder head studs using the double nut method. (Photo).
2. Note that the two longer studs must be fitted to the last two holes towards the rear of the block on the right-hand side. (Photo).
3. Fit a new gasket in place (photo). If one side of the gasket is marked 'TOP' it must naturally be fitted with this side facing upwards.
4. Generously lubricate each cylinder with engine oil (photo).
5. Ensure that the cylinder head face is perfectly clean and then lower the cylinder head into place, keeping it parallel to the block to avoid binding on any of the studs.
6. With the head in place fit the lifting eye over the two rear right-hand studs and the accelerator cable attachment bracket to the next stud along (photo).
7. Fit the cylinder head nuts and washers and tighten down the nuts half a turn at a time in the order shown in Fig. 1.3. to a torque of 45 lb/ft. (Photo).
8. Insert the pushrods into the block so the ball end rests in the tappet. Ensure the pushrods are replaced in the same order in which they were removed. (Photo).
9. Then refit the rocker shaft ensuring that the rocker arm ball joints seat in the pushrod cups. (Photo).
10. Replace the four rocker pedestal nuts and washers and tighten them down evenly. (Photo).

CHAPTER ONE

58.10

59. ROCKER ARM/VALVE ADJUSTMENT

1. The valve adjustments should be made with the engine cold. The importance of correct rocker arm/valve stem clearances cannot be overstressed as they vitally affect the performance of the engine.
2. If the clearances are set too open, the efficiency of the engine is reduced as the valves open late and close earlier than was intended. If, on the other hand the clearances are set too close there is a danger that the stems will expand upon heating and not allow the valves to close properly which will cause burning of the valve head and seat and possible warping.
3. If the engine is in the car to get at the rockers it is merely necessary to remove the two holding down studs from the rocker cover, and then to lift the rocker cover and gasket away.
4. It is important that the clearance is set when the tappet of the valve being adjusted is on the heel of the cam, (i.e. opposite the peak). This can be done by carrying out the adjustments in the following order, which also avoids turning the crankshaft more than necessary.

Valve fully open	Check & adjust
Valve No. 8	Valve No. 1
" " 6	" " 3
" " 4	" " 5
" " 7	" " 2
" " 1	" " 8
" " 3	" " 6
" " 5	" " 4
" " 2	" " 7

5. The correct valve clearance of .010 in. is obtained by slackening the hexagon locknut with a spanner while holding the ball pin against rotation with the screwdriver, (photo). Then, still pressing

59.5

down with the screwdriver, insert a feeler gauge in the gap between the valve stem head and the rocker arm and adjust the ball pin until the feeler gauge will just move in and out without nipping, and, still holding the ball pin in the correct position, tighten the locknut.
6. An alternative method is to set the gaps with the engine running, and although this may be faster it is no more reliable.

60. DISTRIBUTOR & DISTRIBUTOR DRIVE REPLACEMENT

It is important to set the distributor drive correctly as otherwise the ignition timing will be totally incorrect. It is easy to set the distributor drive in apparently the right position, but, exactly 180° out by omitting to select the correct cylinder which must not only be at T.D.C. but must also be on its firing stroke with both valves closed. The distributor drive should therefore not be fitted until the cylinder head is in position and the valves can be observed. Alternatively, if the timing cover has not been replaced, the distributor drive can be replaced when the marks on the timing wheels are adjacent to each other.

1. Rotate the crankshaft so that No.1 piston is at T.D.C. and on its firing stroke (the marks in the timing gears will be adjacent to each other). When No.1 piston is at T.D.C. the inlet valve on No.4 cylinder is just opening and the exhaust valve closing.
2. When the dimple on the crankshaft pulley wheel is in line with the pointer on the timing gear cover, then Nos.1 and 4 pistons are at T.D.C. (photo).

60.2

3. Insert the distributor drive into its housing so that when fully home the slot in the top of the drive shaft is positioned with the larger segment facing downwards in the exact position as shown in the

60.3

ENGINE

photo. The end of the shaft engages with a slot in the top of the pump rotor shaft. It may be necessary to turn the pump rotor shaft to allow the distributor drive to engage fully.

4. It is essential that between 0.002 in. and 0.007 in. endfloat exists between the top side of the gear driven by the skew gear on the camshaft and the underside of the pedestal boss. If the same components are being used it will be safe to assume that the endfloat is correct but ensure the same number of packing washers are used (if any), and always fit a new gasket.

5. If the drive gears are assembled without endfloat wear on the crankshaft gearwheels, chain and distributor drive gear will be very heavy. If new components are being fitted then cut a small notch in the outer edge of the distributor housing flange gasket and bolt the housing down firmly. Measure the thickness of the gasket with a feeler gauge placed in the notch. Then remove the distributor housing and gasket and replace the housing without the gasket. Measure the gap between the underside of the housing flange and the block (photo), and subtract this latter figure from the former to determine the endfloat with the standard gasket.

6. Turn the distributor so the rotor arm is pointing to the terminal in the cap which carries the lead to No.1 cylinder, and fit the distributor to the distributor housing. The lip on the distributor should mate perfectly with the slot in the distributor drive shaft. Fit the bolt which holds the distributor clamp plate to the housing.

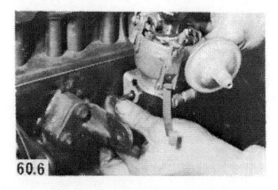

7. Tighten down the two nuts and washers which hold the distributor housing in place.

8. If the clamp bolt on the clamping plate was not previously loosened and the distributor body was not turned in the clamping plate, then the ignition timing will be as previously. If the clamping bolt has been loosened, then it will be necessary to retime the ignition as described in Chapter 4/10.

61. FINAL ASSEMBLY

1. Fit a new gasket to the rocker cover and carefully fit the cover in place. (Photo).

2. Replace the washers over the rocker cover holding down studs ensuring the sealing washer lies under the flat steel washer (photo). Replace the rocker cover nuts.

3. Fit your new sparking plugs (photo). Reconnect the ancilliary components to the engine in the reverse order to which they were removed.

4. It should be noted that in all cases it is best to reassemble the engine as far as possible before refitting it to the car. This means that the inlet and exhaust manifolds, starter motor, water thermostat, oil filter, distributor, carburetters and dynamo, should all be in position. If the engine was removed with the gearbox, the clutch assembly and gearbox

must also be fitted together with the slave cylinder.

62. ENGINE REPLACEMENT

Although the engine can be replaced with one man and a suitable winch, it is easier if two are present. One to lower the engine into the engine compartment and the other to guide the engine into position and to ensure that it does not foul anything. Generally speaking, engine replacement is a reversal of the procedures used when removing the engine, (see Sections 6 and 7), but one or two added tips may come in useful.

1. Ensure all the loose leads, cables etc., are tucked out of the way. If not it is easy to trap one and so cause much additional work after the engine is replaced.
2. Fit the starter motor and oil filter before lowering the engine and gearbox into place.
3. After the dynamo has been replaced it is advisable to fit a new fan belt.
4. Carefully lower the engine into position (photo), and then refit the following:-
a) Gearbox mounting nuts and washers.
b) Front mountings.
c) Propeller shaft to gearbox.
d) Reconnect the clutch pipe to the master cylinder.
e) Speedometer cable.
f) Gearchange remote control and lever (solenoid wires if fitted).
g) Gearbox cover and carpets etc, (refill gearbox first).
h) Oil pressure switch.
i) Rev. counter drive (if fitted).
j) Wires to coil, distributor, and dynamo.
k) Carburetter controls.
l) Fuel pipe to pump and carburetters.
m) Air cleaner/s.
n) Exhaust manifold to pipe and bracket.
o) Earth and starter motor cables.
p) Radiator and hoses and any items hung on radiator attachment bolts.
q) Heater hoses.
r) Water temperature cable.
s) Vacuum advance and retard pipe.
t) Battery.

5. Finally, check that the drain taps are closed and refill the cooling system with water and the engine with the correct grade of oil. Prime the carburetter by working the fuel pump manually, pull out the choke, and start the engine (it should fire first time). Carefully check for oil or water leaks. There should be no oil or water leaks if the engine has been re-assembled carefully, all nuts and bolts tightened down correctly, and new gaskets and joints used throughout.

6. After 500 miles check the tightness of the cylinder head nuts with a torque wrench and change the oil and the filter.

ENGINE FAULT FINDING CHART

Cause	Trouble	Remedy
SYMPTOM:	ENGINE MISFIRES OR IDLES UNEVENLY	
Intermittent sparking at sparking plug	Ignition leads loose	Check and tighten as necessary at spark plug and distributor cap ends.
	Battery leads loose on terminals	Check and tighten terminal leads.
	Battery earth strap loose on body attachment point	Check and tighten earth lead to body attachment point.
	Engine earth lead loose	Tighten lead.
	Low tension leads to SW and CB terminals on coil loose	Check and tighten leads if found loose.
	Low tension lead from CB terminal side to distributor loose	Check and tighten if found loose.
	Dirty, or incorrectly gapped plugs	Remove, clean, and regap.
	Dirty, incorrectly set, or pitted contact breaker points	Clean, file smooth, and adjust.
	Tracking across inside of distributor cover	Remove and fit new cover.
	Ignition too retarded	Check and adjust ignition timing.
	Faulty coil	Remove and fit new coil.
Fuel shortage at engine	Mixture too weak	Check jets, float chamber needle valve, and filters for obstruction. Clean as necessary. Carburettor(s) incorrectly adjusted.
	Air leak in carburettor(s)	Remove and overhaul carburettor.
	Air leak at inlet manifold to cylinder head, or inlet manifold to carburettor	Test by pouring oil along joints. Bubbles indicate leak. Renew manifold gasket as appropriate.
Mechanical wear	Incorrect valve clearances	Adjust rocker arms to take up wear.
	Burnt out exhaust valves	Remove cylinder head and renew defective valves.
	Sticking or leaking valves	Remove cylinder head, clean, check and renew valves as necessary.
	Weak or broken valve springs	Check and renew as necessary.
	Worn valve guides or stems	Renew valve guides and valves.
	Worn pistons and piston rings	Dismantle engine, renew pistons and rings.
SYMPTOM:	LACK OF POWER & POOR COMPRESSION	
Fuel/air mixture leaking from cylinder	Burnt out exhaust valves	Remove cylinder head, renew defective valves.
	Sticking or leaking valves	Remove cylinder head, clean, check, and renew valves as necessary.
	Worn valve guides and stems	Remove cylinder head and renew valves and valve guides.
	Weak or broken valve springs	Remove cylinder head, renew defective springs.
	Blown cylinder head gasket (Accompanied by increase in noise)	Remove cylinder head and fit new gasket.
	Worn pistons and piston rings	Dismantle engine, renew pistons and rings.
	Worn or scored cylinder bores	Dismantle engine, rebore, renew pistons & rings.
Incorrect Adjustments	Ignition timing wrongly set. Too advanced or retarded	Check and reset ignition timing.
	Contact breaker points incorrectly gapped	Check and reset contact breaker points.
	Incorrect valve clearances	Check and reset rocker arm to valve stem gap.
	Incorrectly set sparking plugs	Remove, clean and regap.
	Carburation too rich or too weak	Tune carburettor(s) for optimum performance.
Carburation and ignition faults	Dirty contact breaker points	Remove, clean, and replace.
	Fuel filters blocked causing top end fuel starvation	Dismantle, inspect, clean, and replace all fuel filters.
	Distributor automatic balance weights or vacuum advance and retard mechanisms not functioning correctly	Overhaul distributor.
	Faulty fuel pump giving top end fuel starvation	Remove, overhaul, or fit exchange reconditioned fuel pump.

FAULT FINDING CHART

Cause	Trouble	Remedy
SYMPTOM:	ENGINE FAILS TO TURN OVER WHEN STARTER BUTTON PULLED	
No current at starter motor	Flat or defective battery	Charge or replace battery. Push-start car.
	Loose battery leads	Tighten both terminals and earth ends of earth lead.
	Defective starter solenoid or switch or broken wiring	Run a wire direct from the battery to the starter motor or by-pass the solenoid.
	Engine earth strap disconnected	Check and retighten strap.
Current at starter motor	Jammed starter motor drive pinion	Place car in gear and rock from side to side. Alternatively, free exposed square end of shaft with spanner.
	Defective starter motor	Remove and recondition.
SYMPTOM:	ENGINE TURNS OVER BUT WILL NOT START	
No spark at sparking plug	Ignition damp or wet	Wipe dry the distributor cap and ignition leads.
	Ignition leads to spark plugs loose	Check and tighten at both spark plug and distributor cap ends.
	Shorted or disconnected low tension leads	Check the wiring on the CB and SW terminals of the coil and to the distributor.
	Dirty, incorrectly set, or pitted contact breaker points	Clean, file smooth, and adjust.
	Faulty condenser	Check contact breaker points for arcing, remove and fit new.
	Defective ignition switch	By-pass switch with wire.
	Ignition leads connected wrong way round	Remove and replace leads to spark plugs in correct order.
	Faulty coil	Remove and fit new coil.
	Contact breaker point spring earthed or broken	Check spring is not touching metal part of distributor. Check insulator washers are correctly placed. Renew points if the spring is broken.
No fuel at carburettor float chamber or at jets	No petrol in petrol tank	Refill tank!
	Vapour lock in fuel line (In hot conditions or at high altitude)	Blow into petrol tank, allow engine to cool, or apply a cold wet rag to the fuel line.
	Blocked float chamber needle valve	Remove, clean, and replace.
	Fuel pump filter blocked	Remove, clean, and replace.
	Choked or blocked carburettor jets	Dismantle and clean.
	Faulty fuel pump	Remove, overhaul, and replace. Check CB points on S.U. pumps.
Excess of petrol in cylinder or carburettor flooding	Too much choke allowing too rich a mixture to wet plugs	Remove and dry sparking plugs or with wide open throttle, push-start the car.
	Float damaged or leaking or needle not seating	Remove, examine, clean and replace float and needle valve as necessary.
	Float lever incorrectly adjusted	Remove and adjust correctly.
SYMPTOM:	ENGINE STALLS & WILL NOT START	
No spark at sparking plug	Ignition failure - Sudden	Check over low and high tension circuits for breaks in wiring
	Ignition failure - Misfiring precludes total stoppage	Check contact breaker points, clean and adjust. Renew condenser if faulty.
	Ignition failure - In severe rain or after traversing water splash	Dry out ignition leads and distributor cap.
No fuel at jets	No petrol in petrol tank	Refill tank.
	Petrol tank breather choked	Remove petrol cap and clean out breather hole or pipe.
	Sudden obstruction in carburettor(s)	Check jets, filter, and needle valve in float chamber for blockage
	Water in fuel system	Drain tank and blow out fuel lines

ENGINE

Cause	Trouble	Remedy
SYMPTOM:	EXCESSIVE OIL CONSUMPTION	
Oil being burnt by engine	Badly worn, perished or missing valve stem oil seals Excessively worn valve stems and valve guides Worn piston rings Worn pistons and cylinder bores Excessive piston ring gap allowing blow-by Piston oil return holes choked	Remove, fit new oil seals to valve stems. Remove cylinder head and fit new valves and valve guides. Fit oil control rings to existing pistons or purchase new pistons. Fit new pistons and rings, rebore cylinders. Fit new piston rings and set gap correctly. Decarbonise engine and pistons.
Oil being lost due to leaks	Leaking oil filter gasket Leaking rocker cover gasket Leaking tappet chest gasket Leaking timing case gasket Leaking sump gasket Loose sump plug	Inspect and fit new gasket as necessary. " " " " " " " " " " " " " " " " " " " " " " " " " " " " Tighten, fit new gasket if necessary.
SYMPTOM:	UNUSUAL NOISES FROM ENGINE	
Excessive clearances due to mechanical wear	Worn valve gear (Noisy tapping from rocker box) Worn big end bearing (Regular heavy knocking) Worn timing chain and gears (Rattling from front of engine) Worn main bearings (Rumbling and vibration) Worn crankshaft (Knocking, rumbling and vibration)	Inspect and renew rocker shaft, rocker arms, and ball pins as necessary. Drop sump, if bearings broken up clean out oil pump and oilways, fit new bearings. If bearings not broken but worn fit bearing shells. Remove timing cover, fit new timing wheels and timing chain. Drop sump, remove crankshaft, if bearings worn but not broken up, renew. If broken up strip oil pump and clean out oilways. Regrind crankshaft, fit new main and big end bearings.

CHAPTER TWO

COOLING SYSTEM

CONTENTS

General Description	1
Routine Maintenance	2
Cooling System - Draining	3
Cooling System - Flushing	4
Cooling System - Filling	5
Radiator Removal, Inspection, Cleaning & Replacement	6
Thermostat Removal, Testing & Replacement	7
Water Pump - Removal & Replacement	8
Water Pump - Dismantling & Reassembly	9
Anti-Freeze Mixture	10
Temperature Gauge - Fault Finding	11
Temperature Gauge & Sender Unit - Removal & Replacement	12
Fan Belt Adjustment	13
Fan Belt - Removal & Replacement	14

SPECIFICATIONS

Type of system	Pressurised pump impellor and fan assisted
Thermostat setting - Standard	73°C (163°F)
- Optional	80°C (179°F)

Pressure Cap Opens

948 c.c. Heralds	7 lb.
Herald 1200 & 12/50	7 lb.
from engine No. GA.240782.E.	13 lb.
Herald 13/60	7 lb.
from engine No. GE.22521.E.	13 lb.

Fan Blades

948 c.c. Heralds	2 blades
All other Heralds	4 blades
Tension of fan belt	¾ in. movement midway between dynamo and crankshaft pulley wheels
Type of water pump	Centrifugal
Water pump drive	Belt from crankshaft pulley

Cooling System Capacity

With heater	8.5 pints (10.2 U.S. pints. 4.8 litres)

TORQUE WRENCH SETTINGS

Water outlet elbow nuts	16 to 18 lbs/ft. (2.212 to 2.489 kg.m.)
Water pump to cylinder head	18 to 20 lbs/ft. (2.489 to 2.765 kg.m.)
Water pump pulley attachment	14 to 16 lbs/ft. (1.936 to 2.212 kg.m.)

1. GENERAL DESCRIPTION

The engine cooling water is circulated by a thermo-siphon, water pump assisted, system and the coolant is pressurised. This is to both prevent the loss of water down the overflow pipe with the radiator cap in position and to prevent premature boiling in adverse conditions.

The radiator cap is pressurised and increases the boiling point to 225°F. If the water temperature exceeds this figure and the water boils, the pressure in the system forces the internal part of the cap off its seat, thus exposing the overflow pipe down which

COOLING SYSTEM

the steam from the boiling water escapes thus relieving the pressure.

It is, therefore, important to check that the radiator cap is in good condition and that the spring behind the sealing washer has not weakened. Most garages have a special machine in which radiator caps can be tested.

The cooling system comprises the radiator, top and bottom water hoses, heater hoses (if heater/demister fitted), the impellor water pump, (mounted on the front of the engine it carries the fan blades and is driven by the fan belt), the thermostat and the two drain taps. On 1296 c.c. models the carburetter inlet manifold is heated by water from the pump.

The system functions in the following fashion. Cold water in the bottom of the radiator circulates up the lower radiator hose to the water pump where it is pushed round the water passages in the cylinder block, helping to keep the cylinder bores and pistons cool.

The water then travels up into the cylinder head and circulates round the combustion spaces and valve seats absorbing more heat, and then, when the engine is at its proper operating temperature, travels out of the cylinder head, past the open thermostat into the upper radiator hose and so into the radiator header tank.

The water travels down the radiator where it is rapidly cooled by the in-rush of cold air through the radiator core, which is created by both the fan and the motion of the car. The water, now cold, reaches the bottom of the radiator, when the cycle is repeated.

When the engine is cold the thermostat (which is a valve which opens and closes according to the temperature of the water) maintains the circulation of the same water in the engine.

Only when the correct minimum operating temperature has been reached, as shown in the specification, does the thermostat begin to open, allowing water to return to the radiator.

2. ROUTINE MAINTENANCE

1. Check the level of the water in the radiator once a week or more frequently if necessary, and top up with a soft water (rain water is excellent) as required.
2. Once every 6,000 miles check the fan belt for wear and correct tension and renew or adjust the belt as necessary. (See Sections 13 and 14 for details).
3. Once every 12,000 miles unscrew the plug from the top of the water pump, fit a grease nipple and give five strokes with the grease gun supplied. On later models a grease nipple is fitted as standard. Do not overgrease or the seal may be rendered inoperative. Replace the plug and screw down.

3. COOLING SYSTEM - DRAINING

1. With the car on level ground drain the system as follows:-
2. If the engine is cold remove the filler cap from the radiator by turning the cap anti-clockwise. If the engine is hot having just been run, then turn the filler cap very slightly until the pressure in the system has had time to disperse. Use a rag over the cap to protect your hand from escaping steam. If, with the engine very hot, the cap is released suddenly, the drop in pressure can result in the water boiling. With the pressure released the cap can be removed.
3. If anti-freeze is in the radiator drain it into a clean bucket or bowl for re-use.
4. Open the two drain taps and ensure the heater control is in the hot position. On later models drain plugs may be fitted instead of taps. Remove the plugs with a spanner. The drain taps/plugs are located at the bottom of the radiator and at the rear on the right-hand side of the block (photo).

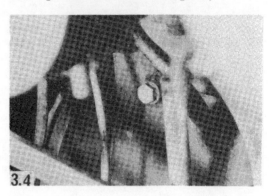
3.4

5. When the water has finished running, probe the drain tap orifices with a short piece of wire to dislodge any particles or rust or sediment which may be blocking the taps and preventing all the water draining out.

4. COOLING SYSTEM - FLUSHING

1. With time the cooling system will gradually lose its efficiency as the radiator becomes choked with rust scales, deposits from the water and other sediment. To clear the system out, remove the radiator cap and the drain taps and leave a hose running in the radiator cap orifice for ten to fifteen minutes.
2. Then close the drain taps and refill with water and a proprietary cleansing compound. Run the engine for 10 to 15 minutes and then drain it and flush out thoroughly for a further ten minutes. All sediment and sludge should now have been removed.
3. In very bad cases the radiator should be reverse flushed. This can be done with the radiator in position. The cylinder block tap is closed and a hose placed over the open radiator drain tap. Water, under pressure, is then forced up through the radiator and out of the header tank filler orifice.
4. The hose is then removed and placed in the filler orifice and the radiator washed out in the usual fashion.

COOLING SYSTEM - FILLING

1. Close the two drain taps.
2. Fill the system slowly to ensure that no air locks develop. If a heater is fitted, check that the valve to the heater unit is open, otherwise an air lock may form in the heater. The best type of water to use in the cooling system is rain water, so use this whenever possible.

55

CHAPTER TWO

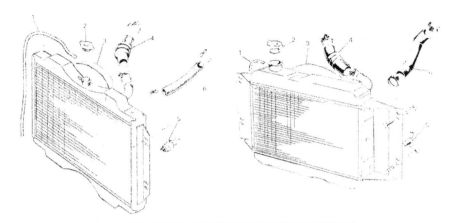

Fig. 2.1. TWO TYPES OF RADIATOR FITTED TO HERALD
1 Overflow pipe. 2 Filler cap. 3 Radiator. 4 Top hose(s). 5 Drain circle. 6 Bottom hose. 7 Overflow reservoir. 8 Header tank.

3. Do not fill the system higher than within ½ in. of the filler orifice. Overfilling will merely result in wastage, which is especially to be avoided when anti-freeze is in use.
4. Only use anti-freeze mixture with a glycerine or ethylene base.
5. Replace the filler cap and turn it firmly clockwise to lock it in position.

6. RADIATOR REMOVAL, INSPECTION, CLEANING & REPLACEMENT
1. To remove the radiator first drain the cooling system as described in Section 2.
2. Then undo the jubilee clip which holds the top water hose to the thermostat pipe outlet. Disconnect the horn wires at the snap connector where the horns are hung on the radiator securing bolts.
3. Pull the top water hose off the thermostat elbow, and then undo the jubilee clip on the bottom hose and pull the end of the hose off the radiator.
4. Undo and remove the bolts, and washers which hold the radiator in place. On later models the radiator is attached to side valances. Unhook the bonnet tension springs from their anchor brackets and undo the bolts holding each side of the radiator to the valance.
5. Lift the radiator up out of the engine compartment.
6. With the radiator out of the car any leaks can be soldered up or repaired with a substance such as 'cataloy'. Clean out the inside of the radiator by flushing as detailed in the section before last. When the radiator is out of the car it is advantageous to turn it upside down for reverse flushing. Clean the exterior of the radiator by hosing down the radiator matrix with a strong jet of water to clear away road dirt, dead flies etc.
7. Inspect the radiator hoses for cracks, internal or external perishing, and damage caused by overtightening of the securing clips. Replace the hoses as necessary. Examine the radiator hose securing clips and renew them if they are rusted or distorted. The drain taps should be renewed if leaking, but ensure the leak is not because of a faulty washer behind the tap. If the tap is suspected try a new washer to see if this clears the touble first.

8. Replacement is a straightforward reversal of the removal procedure.

7. THERMOSTAT REMOVAL, TESTING & REPLACEMENT
1. To remove the thermostat partially drain the cooling system (4 pints is enough), loosen the upper radiator hose at the thermostat elbow end and pull it off the elbow. Disconnect the wire from the Lucar connector on the sender unit on models where the sender unit is in the thermostat cover.

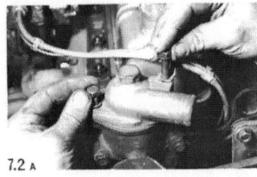

7.2 A

2. Unscrew the two set bolts and spring washers from the thermostat housing and lift the housing and paper gasket away. Take out the thermostat.

7.2 B

3. Test the thermostat for correct functioning by dangling it by a length of string in a saucepan of cold water together with a thermometer.
4. Heat the water and note when the thermostat

COOLING SYSTEM

begins to open. This temperature is stamped on the flange of the thermostat, and is also given in the specifications on page 54.

5. Discard the thermostat if it opens too early. Continue heating the water until the thermostat is fully open. Then let it cool down naturally. If the thermostat will not open fully in boiling water, or does not close down as the water cools, then it must be exchanged for a new one.

6. If the thermostat is stuck open when cold this will be apparent when removing it from the housing.

7. Replacing the thermostat is a reversal of the removal procedure. Remember to use a new paper

gasket between the thermostat housing elbow and the thermostat. Renew the thermostat elbow if it is badly eaten away.

8. **WATER PUMP - REMOVAL & REPLACEMENT**

1. If the water pump is badly worn normal practice is to fit an exchange reconditioned unit. Drain the cooling system as described in Section 2 and slacken the dynamo mounting bolts so the fan belt can be removed.

2. Undo the clips which hold the top and bottom water hoses to the pump body. On early models fitted with a heater, free the return pipe by undoing the union nut on the rear of the pump body. On later models with a heated inlet manifold free the small hose from the pump which leads to the manifold.

3. Pull off the temperature transmitter wire (where fitted) from its Lucar connector.

4. Undo the three bolts and washers which hold the pump body to the front of the cylinder block. Note that the bolts are all of different length. Lift the combined fan and water pump away from the engine

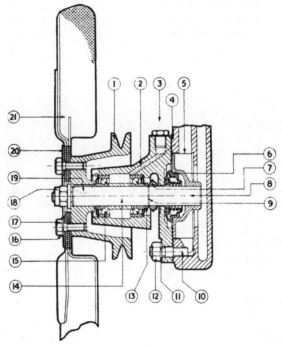

Fig. 2.2. SECTIONED VIEW OF WATER PUMP
1 Fan pulley. 2 Bearing housing. 3 Plug. 4 Water pump body. 5 Impellor. 6 Sealing gland. 7 Rubber ring. 8 Spindle. 9 Circlip. 10 Stud. 11 Spring washer. 12 Nut. 13 Washer. 14 Distance piece. 15 Bearings. 16 Spring washer. 17 Bolt. 18 Nyloc nut. 19 Woodruff key. 20 Balancer. 21 Fan blade assembly.

CHAPTER TWO

and remove the gasket.

5. Replacement is a straightforward reversal of the removal sequence. Note that the fan belt tension must be correct when all is reassembled. If the belt is too tight undue strain will be placed on the water pump and dynamo bearings, and if the belt is too loose it will slip and wear rapidly as well as giving rise to low electrical output from the dynamo.

9. WATER PUMP - DISMANTLING & REASSEMBLY

1. If it is wished to repair the pump first ascertain that spare parts are available, as less and less firms stock spare parts as opposed to rebuilt pump units. Remove the four bolts and spring washers which hold the fan blade (25, 26) in place and carefully note the position of the balance weight (27) if fitted. (References are to Fig. 2.4.).

2. Undo the nut and washer (16, 17) which hold the fan pulley (14) in place and with the aid of an extractor pull off the pulley wheel and prise out the woodruff key (15).

3. Undo the three nuts and spring washers (19, 20) which hold the bearing housing (3) to the pump body (1) and pull out the bearing housing.

4. With the aid of a vice and an extractor pull the impellor (7) off the spindle (5). Remove the sealing gland (6) from the back of the impellor (7).

5. Remove the bearing retaining circlip (13) from the bore of the housing and pull out the spindle complete with bearings. The bearings (11) distance piece (12), circlip (10), washer (9) and bearing seal (8) can now all be removed.

6. If the pump is badly worn the bearings will require renewal and the gland face on the housing recut. (This is a job for a TRIUMPH garage or your local engineering works). A new sealing gland and bearing seal, together with a new gasket must also be obtained.

7. Reassembly of the water pump is a reversal of the above sequence. The following additional points should be noted.

8. Position the bearings so that their unshielded sides are adjacent to the distance piece and the grease seal faces outwards. Pack the bearings and area round the distance piece with grease.

9. The shaft and bearings are fitted to the housing with the aid of a drift made from a piece of tubing.

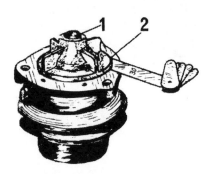

Fig. 2.3. WATER PUMP IMPELLOR ASSEMBLED TO SPINDLE
1 Solder. 2 Feeler gauge measuring .030 in. clearance between impellor and pump body.

10. Press the impellor onto the spindle until a 0.030 in. clearance measured with a feeler gauge exists between the flat face of the impellor and the housing. The impellor should then be soldered to the shaft to prevent water seepage down the spindle.

11. When fitting the pulley wheel, fan and balance weight, note if a small alignment hole has been drilled in these units. If so line the components up with the aid of a $1/16$ in. drill or similar while the securing bolts are being done up.

12. Regrease the water pump on completion of the assembling operations.

10. ANTI-FREEZE MIXTURE

1. In circumstances where it is likely that the temperature will drop to below freezing it is essential that some of the water is drained and an adequate amount of ethylene glycol antifreeze such as Bluecol added to the cooling system.

2. If Bluecol is not available any antifreeze which conforms with specification B.S. 3151 or B.S. 3152 can be used. Never use an antifreeze with an alcohol base as evaporation is too high.

3. Bluecol antifreeze with an anti-corrosion additive can be left in the cooling system for up to two years, but after six months it is advisable to have the specific gravity of the coolant checked at your local garage, and thereafter once every three months.

4. Listed below are the amounts of Bluecol which should be added to ensure adequate protection down to the temperature given.

Amount of A.F.	Protection to
1.7 pints (1 litre)	$-17.8^{\circ}C$ ($0^{\circ}F$)
2.0 pints (1.3 litres)	$-28.9^{\circ}C$ ($-20^{\circ}F$)
2.3 pints (1.43 litres)	$-34.5^{\circ}C$ ($-30^{\circ}F$)
3.0 pints (1.7 litres)	$-40^{\circ}C$ ($-40^{\circ}F$)

11. TEMPERATURE GAUGE - FAULT FINDING

1. If the temperature gauge fails to work either the gauge, the sender unit, the wiring or the connections are at fault.

2. It is not possible to repair the gauge or the sender unit and they must be replaced by new units if at fault.

3. First check the wiring connections and if sound check the wiring for breaks using an ohmmeter. The sender unit and gauge should be tested by substitution.

12. TEMPERATURE GAUGE & SENDER UNIT - REMOVAL & REPLACEMENT

1. For details of how to remove and replace the temperature gauge see Chapter 12, Section 12.

2. To remove the sender unit disconnect the battery, pull off the wire at the snap connector on the unit, and undo the unit with a spanner. On replacement renew the fibre washer to prevent the possibility of leaks developing.

13. FAN BELT ADJUSTMENT

1. It is important to keep the fan belt correctly adjusted and although not listed by the manufacturer, it is considered that this should be a regular maintenance task performed every 6,000 miles.

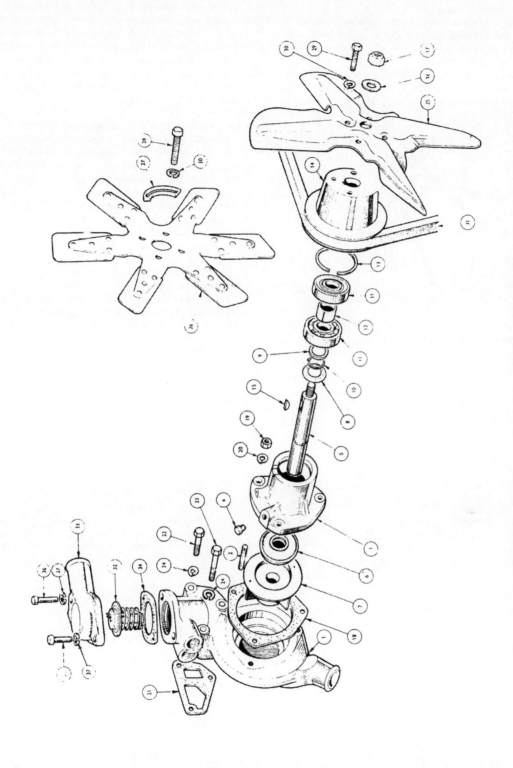

Fig. 2.4. EXPLODED VIEW OF THE WATER PUMP, FAN ASSEMBLY, THERMOSTAT AND WATER OUTLET ELBOW

1 Water pump body. 2 Stud. 3 Water pump bearing housing. 4 Grease plug. 5 Spindle. 6 Seal. 7 Impellor. 8 Bearing seal. 9 Abutment washer. 10 Circlip. 11 Bearings. 12 Distance piece. 13 Bearing retaining circlip. 14 Fan pulley. 15 Woodruff key. 16 Retaining washer. 17 Nyloc nut. 18 Gasket. 19 Nut. 20 Lock washer. 21 Gasket. 22 Short bolt (1). 23 Intermediate length bolt (1) and long bolt (1) (only one bolt shown. 24 Lock washer. 25 Normal fan. 26 Special fan (certain overseas markets only). 27 Balance piece. 28 Short bolt (1). 29 Bolt for normal fan (4). 30 Spring washers ('). 31 Fan belt. 32 Thermostat unit. 33 Water outlet elbow. 34 Gasket. 35 Bolt. 36 Longer bolt. 37 Spring washers.

2. If the belt is too loose it will slip, wear rapidly, and cause the dynamo and water pump to malfunction. If the belt is too tight the dynamo and water pump bearings will wear rapidly causing premature failure of these components.

3. The fan belt tension is correct when there is ¾ in. of lateral movement at the midpoint position of the belt between the dynamo pulley wheel and the crankshaft pulley wheel.

4. To adjust the fan belt, slacken the dynamo securing bolts and move the dynamo either in or out until the correct tension is obtained. It is easier if the dynamo bolts are only slackened a little so it requires some force to move the dynamo. In this way the tension of the belt can be arrived at more quickly than by making frequent adjustments.

5. With the dynamo bolts only slightly loosened, difficulty may be experienced in moving the dynamo away from the engine. A long spanner placed behind the dynamo and resting against the block serves as a very good lever and can be held in this position while the dynamo bolts are tightened. On no account overtighten the fan belt. It is better for the belt to be too loose than too tight.

14. FAN BELT - REMOVAL & REPLACEMENT

1. If the fan belt is worn or has stretched unduly it should be replaced. The most usual reason for replacement is that the belt has broken in service. It is therefore recommended that a spare belt is always carried. Replacement is a reversal of the removal sequence, but as replacement due to breakage is the most usual operation, it is described below.

2. Loosen the two dynamo pivot bolts and the nut on the adjusting link and push the dynamo in towards the engine.

3. Slip the belt over the crankshaft, dynamo, and water pump pulleys.

4. Adjust the belt as described in the previous section and tighten the dynamo mounting nuts. NOTE after fitting a new belt it will require adjustment 250 miles later.

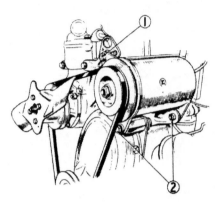

Fig. 2.5. FAN BELT ADJUSTMENT
1 Generator adjusting link bolt. 2 Generator pivot bolts.

COOLING SYSTEM
FAULT FINDING CHART

Cause	Trouble	Remedy
SYMPTOM:	**OVERHEATING**	
Heat generated in cylinder not being successfully disposed of by radiator	Insufficient water in cooling system	Top up radiator
	Fan belt slipping (Accompanied by a shrieking noise on rapid engine acceleration	Tighten fan belt to recommended tension or replace if worn.
	Radiator core blocked or radiator grill restricted	Reverse flush radiator, remove obstructions.
	Bottom water hose collapsed, impeding flow	Remove and fit new hose.
	Thermostat not opening properly	Remove and fit new thermostat.
	Ignition advance and retard incorrectly set (Accompanied by loss of power, and perhaps, misfiring)	Check and reset ignition timing.
	Carburettor(s) incorrectly adjusted (mixture too weak)	Tune carburettor(s).
	Exhaust system partially blocked	Check exhaust pipe for constrictive dents and blockages.
	Oil level in sump too low	Top up sump to full mark on dipstick.
	Blown cylinder head gasket (Water/steam being forced down the radiator overflow pipe under pressure)	Remove cylinder head, fit new gasket.
	Engine not yet run-in	Run-in slowly and carefully.
	Brakes binding	Check and adjust brakes if necessary.
SYMPTOM:	**UNDERHEATING**	
Too much heat being dispersed by radiator	Thermostat jammed open	Remove and renew thermostat.
	Incorrect grade of thermostat fitted allowing premature opening of valve	Remove and replace with new thermostat which opens at a higher temperature.
	Thermostat missing	Check and fit correct thermostat.
SYMPTOM	**LOSS OF COOLING WATER**	
Leaks in system	Loose clips on water hoses	Check and tighten clips if necessary.
	Top, bottom, or by-pass water hoses perished and leaking	Check and replace any faulty hoses.
	Radiator core leaking	Remove radiator and repair.
	Thermostat gasket leaking	Inspect and renew gasket.
	Radiator pressure cap spring worn or seal ineffective	Renew radiator pressure cap.
	Blown cylinder head gasket (Pressure in system forcing water/steam down overflow pipe	Remove cylinder head and fit new gasket.
	Cylinder wall or head cracked	Dismantle engine, dispatch to engineering works for repair.

CHAPTER THREE

FUEL SYSTEM AND CARBURATION

CONTENTS

General Description...	1
Air Cleaners - Removal, Replacement & Servicing	2
A.C. Fuel Pump - Description	3
A.C. Fuel Pump - Routine Maintenance	4
A.C. Fuel Pump - Removal & Replacement	5
A.C. Fuel Pump - Testing	6
A.C. Fuel Pump - Dismantling	7
A.C. Fuel Pump - Examination & Reassembly	8
S.U. Carburetters - Routine Maintenance	9
S.U. Carburetters - Description	10
S.U. Carburetters - Removal & Replacement	11
S.U. Carburetters - Dismantling	12
S.U. Carburetter Float Chamber - Dismantling, Examination & Reassembly	13
S.U. Carburetter Float Chamber - Fuel Level Adjustment	14
S.U. Carburetter - Examination & Repair	15
S.U. Carburetter - Piston Sticking	16
S.U. Carburetter - Float Needle Sticking	17
S.U. Carburetter - Float Chamber Flooding	18
S.U. Carburetter - Water & Dirt in Carburetter	19
S.U. H1 Carburetter - Jet Centring	20
S.U. Carburetter - Adjustment & Tuning	21
Synchronisation of Twin S.U. Carburetter	22
Solex B28 21C-2 Carburetter - Description	23
Solex B28 21C-2 Carburetter - Removal & Replacement	24
Solex B28 21C-2 Carburetter - Dismantling & Reassembly	25
Solex B28 21C-2 Carburetter - Adjustment	26
Solex B30 PSE1 Carburetter - Removal, Dismantling & Replacement	27
Solex B30 PSE1 Carburetter - Slow Running Adjustment	28
Stromberg 150 CD Carburetter - Description	29
Stromberg 150 CD Carburetter - Adjustments	30
Stromberg 150 CD Carburetter - Float Chamber Fuel Level Adjustment	31
Stromberg 150 CD Carburetter - Dismantling & Reassembly	32
Fuel Tank - Removal & Replacement	33
Fuel Gauge - Removal & Replacement	34
Solex Carburetter - Lack of Fuel at Engine	35
Solex Carburetters - Weak Mixture	36
Solex Carburetters - Rich Mixture	37

SPECIFICATIONS

Fuel Pump
    Make & type ... A.C. mechanically operated diaphragm
    Pump pressure ... $1\frac{1}{2}$ to $2\frac{1}{2}$ lb/sq.in.
    Fitted length of diaphragm spring.. ... .468 in.
    Fitted load ... $4\frac{1}{8}$ to $4\frac{3}{8}$ lb.
    Free length ... $1\frac{1}{8}$ in.

Tank Capacity
    Saloons & Coupes ... 6.5 galls. (7.3 U.S. galls. 32.0 litres)
    Estates & Courier ... 9.0 galls. (10.8 U.S. galls. 41.0 litres)

FUEL SYSTEM AND CARBURATION

Carburetters
  Make & type
    948 c.c. saloons ... ...   Single Solex B28 21C-2
    948 c.c. Coupes ... ...   Twin S. U. H1
    Herald 1200 & 12/50 ... ...   Single Solex B30 PSE1
    Herald 13/60... ... ...   Single 150 CD Stromberg
  Dashpot hydraulic damper oil ... ...   S.A.E. 20

Air Cleaners
  Make & type
    948 c.c. saloons ... ...   A.C. wire gauze or oil bath
    948 c.c. Coupes ... ...   A.C. wire gauze or replaceable paper element
    Herald 1200 & 12/50 ... ...   A.C. replaceable paper element
    Herald 13/60... ... ...   A.C. pancake type replaceable paper element

TORQUE WRENCH SETTING

Fuel pump studs.. ... ...   12 to 14 lbs/ft. (1.659 to 1.936 kg.m.)

1. GENERAL DESCRIPTION

The fuel system on all models of the Herald consists of a fuel tank mounted at the rear of the car either in the left-hand side of the boot or under the boot floor; an A.C. mechanical fuel pump mounted on the left-hand side of the engine and operated by the camshaft; the necessary fuel lines between the tank and the pump; and the pump and the carburetter/s; a single Solex B2821C-2 carburetter on 948 c.c. saloons, twin H 1.S.U. on 948 c.c. Coupes and convertibles, a single Solex B30PSE1 on the Herald 1200 and 12/50, and a single 150 CD Stromberg fitted to the 13/60. A fuel contents gauge is fitted to all models.

2. AIR CLEANERS - REMOVAL, REPLACEMENT & SERVICING

1. Several different types of air cleaner were fitted depending on year of manufacture and type of carburetters used. The cleaners should be serviced every 3,000 miles and where paper element cleaners are used the element should be renewed every 12,000 miles.
2. Early models are fitted with either an oil bath cleaner (1), (Fig. 3.2.) or a gauze type cleaner (2), (Fig. 3.2.). To deal first with the oil bath cleaner every 3,000 miles or whenever the engine oil is changed also renew the air cleaner oil.
3. Undo the two clips (18, 21), and lift away the duct (19). Undo the winged retaining bolt (20), take off the cover (15) and then lift out the element (16) from the container (14).
4. Thoroughly clean the filter element (16) in petrol, clean and wipe out the container and refill with fresh engine oil to the level marked. Replacement is a straightforward reversal of the removal sequence but take special care that the rubber joint (17) is in its correct position and all joints are air-tight.
5. To service the gauze type of cleaner loosen the clip (3), take out the retaining bolt and washer (7, 8) and remove the cleaner (4). Thoroughly wash the cleaner in petrol, re-oil it, let it drain, and then refit it to the engine.
6. Later models are fitted with paper element type air cleaners, and these are serviced fairly easily. Undo the bolt/s which holds the air cleaner cover/s in place, and on some models it is also necessary to release a wire clip and withdraw its ends from the holes in the edge of the casing. (See Fig. 3.2.).
7. Clean the interior of the air cleaner cover, blow and brush out the dirt from the folds of the air-cleaner element. Every 12,000 miles or earlier if the element becomes torn renew the paper filter.
8. On replacement renew the air cleaner/carburetter gaskets if necessary and make sure the holes in the gaskets are correctly aligned and the filter element properly positioned.

3. A.C. FUEL PUMP - DESCRIPTION

The mechanically operated A.C. fuel pump is actuated through spring loaded rocker arm. One arm of the rocker (D) bears against an eccentric (H) on the camshaft (G) and the other arm operates a diaphragm pull rod (F). NOTE all references in brackets should be co-related with Fig. 3.1.

As the engine camshaft rotates, the eccentric moves the pivoted rocker arm outwards which in turn pulls the diaphragm pull rod (F) and the diaphragm (A) down against the pressure of the diaphragm spring (C).

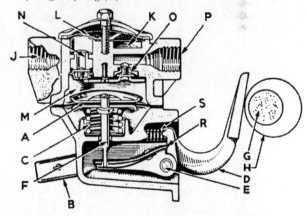

Fig. 3.1. A SECTIONED VIEW OF A.C. FUEL PUMP
A Diaphragm. B Hand priming lever. C Diaphragm spring. D Rocker arm. E Diaphragm operating arm. F Diaphragm pull rod. G Camshaft. H Camshaft eccentric. J Petrol pipe union. L Fuel filter gauze. M Pump chamber. N Non-return valve. O Non-return outlet valve. P Petrol pipe union. R Diaphragm pull rod retainer. S Anti-rattle spring.

This creates sufficient vacuum in the pump chamber (M) to draw in fuel from the tank through the fuel filter gauze (L), and non-return valve (N).

The rocker arm is held in constant contact with the eccentric by an anti-rattle spring (S), and as the engine camshaft continues to rotate the eccentric allows the rocker arm to move inwards. The diaphragm spring (C) is thus free to push the diaphragm (A) upwards forcing the fuel in the pump chamber (M) out to the carburetter through the non-return outlet valve (O).

When the float chamber in the carburetter is full the float chamber needle valve will close so preventing further flow from the fuel pump.

The pressure in the delivery line will hold the diaphragm downwards against the pressure of the diaphragm spring, and it will remain in this position until the needle valve in the float chamber opens to admit more petrol.

4. A.C. FUEL PUMP - ROUTINE MAINTENANCE
1. Every 6,000 miles (3,000 miles in very dusty conditions) undo the bolt in the centre of the cover and lift off the cover.
2. Inspect the filter gauze for sediment and lift it out and clean it with petrol and a soft brush if dirty.
3. Check the condition of the cork gasket and renew if it has hardened or broken. Replacement is a straightforward reversal of the removal sequence - do not forget to refit the fibre washer under the head of the retaining bolt and tighten the bolt just enough for the cover to make a leak-proof joint with the gasket.

5. A.C. FUEL PUMP - REMOVAL & REPLACEMENT
1. Remove the fuel inlet and outlet pipes by unscrewing the union nuts.
2. Undo the two nuts and spring washers which hold the pump to the crankcase. Note the special nut at the rear of the pump with a slotted head (photo). Ensure it is replaced on the correct stud, i.e. the stud nearest the rear of the engine.

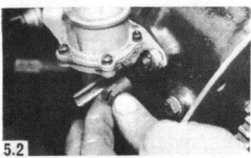

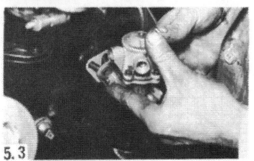

3. Lift the pump together with the gasket away from the crankcase. (Photo).
4. Replacement of the pump is a reversal of the above process. Remember to use a new crankcase to fuel pump gasket to ensure no oil leaks (photo), ensure that both faces of the flange are perfectly clean, and check that the rocker arm lies on top of the camshaft eccentric and not underneath it.

6. A.C. FUEL PUMP TESTING
Presuming that the fuel lines and unions are in good condition and that there are no leaks anywhere, check the performance of the fuel pump in the following manner. Disconnect the fuel pipe at the carburetter inlet union, and the high tension lead to the coil, and with a suitable container or a large rag in position to catch the ejected fuel, turn the engine over on the starter motor solenoid. A good spurt of petrol should emerge from the end of the pipe every second revolution.

7. A.C. FUEL PUMP DISMANTLING
1. Unscrew the securing bolt from the centre of the cover and lift the cover away. NOTE the fibre washer under the head of the bolt.
2. Remove the cork sealing washer and the fine mesh filter gauze.
3. If the condition of the diaphragm is suspect or for any other reason it is wished to dismantle the pump fully, proceed as follows:- Mark the upper and lower flanges of the pump that are adjacent to each other. Unscrew the five screws and spring washers which hold the two halves of the pump body together. Separate the two halves with great care, ensuring that the diaphragm does not stick to either of the two flanges.
4. Unscrew the screws which retain the valve plate and remove the plate and gasket together with the inlet and outlet valves. (Some later pumps have a simplified valve plate arrangement which is released by one screw).
5. Press down and rotate the diaphragm a quarter of a turn (in either direction) to release the pull rod from the operating lever, and lift away the diaphragm and pull rod (which is securely fixed to the diaphragm and cannot be removed from it). Remove the diaphragm spring and the metal and fibre washer underneath it.
6. If it is necessary to dismantle the rocker arm assembly, remove the retaining circlips and washer from the rocker arm pivot rod and slide out the rod which will then free the rocker arm, operating rod, and anti-rattle spring.

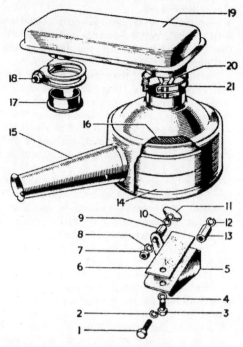
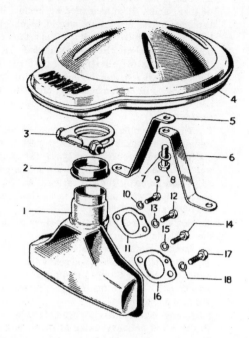

EXPLODED VIEW OF THE OIL BATH TYPE OF AIR CLEANER

EXPLODED VIEW OF THE GAUZE TYPE OF AIR CLEANER

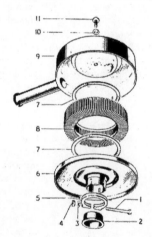

EXPLODED VIEW OF THE 1200 AND 12/50 PAPER ELEMENT AIR CLEANER

EXPLODED VIEW OF THE 13/60 PAPER ELEMENT AIR CLEANER

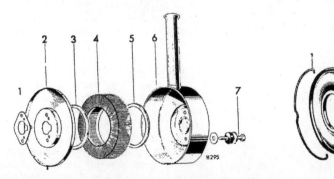

EXPLODED VIEW OF THE PAPER ELEMENT TYPE OF AIR CLEANER

Fig. 3.2. TYPES OF AIR CLEANER

CHAPTER THREE

8. A.C. FUEL PUMP EXAMINATION & REASSEMBLY

1. Check the condition of the cork cover sealing washer, and if it is hardened or broken it must be replaced. The diaphragm should be checked similarly and replaced if faulty. Clean the pump thoroughly and agitate the valves in paraffin to clean them out. This will also improve the contact between the valve seat and the valve. It is unlikely that the pump body will be damaged, but check for fractures and cracks. Renew the cover if distorted by over-tightening.

2. To reassemble the pump proceed as follows:- Replace the rocker arm assembly comprising the operating link, rocker arm, anti-rattle spring and washer in their relative positions in the pump body. Align the holes in the operating link, rocker arm, and washers with the holes in the body and insert the pivot pin.

3. Refit the circlips to the grooves in each end of the pivot pin.

4. Earlier pumps used valves which had to be built up, while later versions used ready assembled valves which are merely dopped into place in the inlet and outlet ports. Ensure that the correct valve is dropped into each port.

5. Reassemble the earlier type of valve as follows:- Position the delivery valve in place on its spring. Place the inlet valve in position in the pump body and then fit the spring. Place the small four legged inlet valve spring retainer over the spring with the legs positioned towards the spring.

6. Place the valve retaining gasket in position, replace the plate, and tighten down the three securing screws. (Or single screw in the case of later models). Check that the valves are working properly with a suitable piece of wire.

7. Position the fibre and steel washer in that order in the base of the pump and place the diaphragm spring over them.

8. Replace the diaphragm and pull rod assembly with the pull rod downwards and the small tab on the diaphragm adjacent to the centre of the flange and rocker arm.

9. With the body of the pump held so that the rocker arm is facing away from one, press down the diaphragm, turning it a quarter of a turn to the left at the same time. This engages the slot on the pull rod with the operating lever. The small tab on the diaphragm should now be at an angle of $90°$ to the rocker arm and the diaphragm should be firmly located.

10. Move the rocker arm until the diaphragm is level with the body flanges and hold the arm in this position. Reassemble the two halves of the pump ensuring that the previously made marks on the flanges are adjacent to each other.

11. Insert the five screws and lockwashers and tighten them down finger tight.

12. Move the rocker arm up and down several times to centralise the diaphragm, and then with the arm held down, tighten the screws securely in a diagonal sequence.

13. Replace the gauze filter in position. Fit the cork cover sealing washer, fit the cover, and insert the bolt with the fibre washer under its head. Do not over-tighten the bolt but ensure that it is tight enough to prevent any leaks.

9. S.U. CARBURETTERS - ROUTINE MAINTENANCE

1. Once every 3,000 miles undo the hexagon caps on the dashpot/s and top them up to within $\frac{1}{2}$ in. of the top with Castrolite or a similar S.A.E. 20 oil as shown under 'recommended lubricants' on page 10.

2. Once every 6,000 miles adjust the carburetter slow running and tune the carburetters if necessary. See sections 21 and 22 for further details. Check the fuel lines and the union joints for leaks or weeping and replace defective washers as required.

3. Also every 6,000 miles remove the float chambers from the carburetters, empty away any sediment, check the condition of the needle valve, clean and reassemble. Remove and clean the filters in the carburetters and fuel pump where these are fitted.

10. S.U. CARBURETTERS - DESCRIPTION

The variable choke S.U. carburetter is a relatively simple instrument and is basically the same irrespective of its size and type. It differs from most other carburetters in that instead of having a number of various sized fixed jets for different conditions, only one variable jet is fitted to deal with all possible conditions.

Air passing rapidly through the carburetter choke draws petrol from the jet so forming the petrol/air mixture. The amount of petrol drawn from the jet depends on the position of the tapered carburetter needle, which moves up and down the jet orifice according to engine load and throttle opening, thus effectively altering the size of the jet so that exactly the right amount of fuel is metered for the prevailing road conditions.

The position of the tapered needle in the jet is

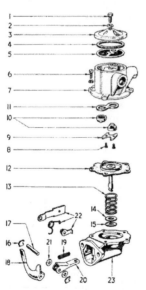

Fig. 3.3. EXPLODED VIEW OF THE A.C. MECHANICAL FUEL PUMP

1 Cover securing screw. 2 Washer. 3 Cover. 4 Cork gasket. 5 Filler gauge. 6 Flange set screw (5 off). 7 Upper pump body. 8 Screws. 9 Retainer. 10 Valves. 11 Upper retainer. 12 Diaphragm assembly. 13 Diaphragm spring. 14 Plain washer. 15 Fabric gland washer. 16 Pivot pin retainer. 17 Pivot pin. 18 Rocker arm. 19 Rocker arm section spring. 20 Operating fork. 21 Distance washer. 22 Hand priming lever assembly. 23 Lower pump body.

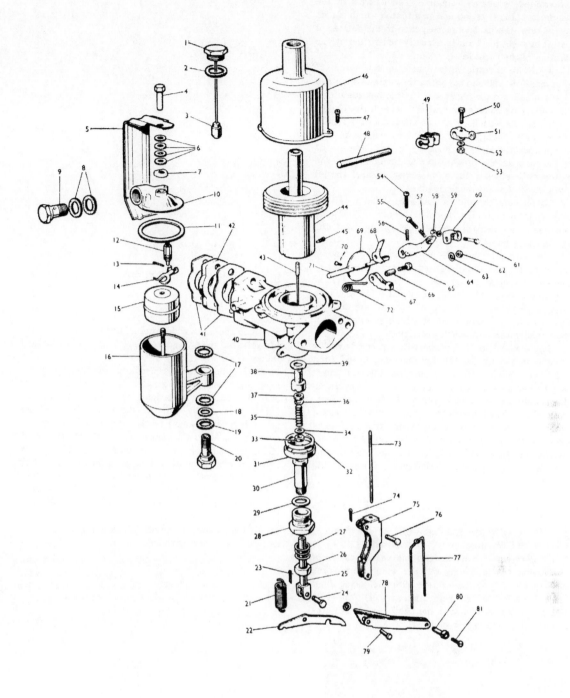

Fig. 3.4. EXPLODED VIEW OF THE S.U. H.1. CARBURETTER

1 Damper nut. 2 Washer. 3 Damper plunger. 4 Cap nut. 5 Heat shield. 6 Spacer washers. 7 Distance piece. 8 Fibre washers. 9 Banjo union. 10 Float chamber lid. 11 Washer. 12 Needle valve. 13 Fulcrum pin. 14 Lever. 15 Float. 16 Float chambers. 17 Fibre washers. 18 Washer. 19 Fibre washer. 20 Union. 21 Return spring. 22 Jet lever. 23 Split pin. 24 Clevis pin. 25 Jet head. 26 Adjusting nut. 27 Spring. 28 Gland nut. 29 Copper washer. 30 Lower jet bearing. 31 Sealing ring. 32 Gland washer. 33 Cork gland washer. 34 Washer. 35 Gland washer spring. 36 Washer. 37 Cork gland washer. 38 Upper jet bearing. 39 Copper washer. 40 Body. 41 Flange points. 42 Insulation washer. 43 Needle. 44 Piston assembly. 45 Screw. 46 Suction chamber (Dashpot). 47 Screw. 48 Throttle connecting rod. 49 Flexible joint. 50 Clamp bolt. 51 Throttle lever. 52 Washer. 53 Nut. 54 Throttle adjusting screws. 55 Mixture adjusting screw. 56 Spring. 57 Spring. 58 Nut. 59 Washer. 60 Coupling lever. 61 Pinch bolt. 62 Nut. 63 Washer. 64 Stop lever. 65 Set screw. 66 Distance piece. 67 Rocker arm. 68 Cam lever. 69 Throttle disc. 70 Screw. 71 Throttle spindle. 72 Spring. 73 Pushrod. 74 Split pin. 75 Jet link. 76 Pivot pin. 77 Stirrup. 78 Jet lever. 79 Clevis pin. 80 Pivot pin. 81 Screw.

determined by engine vacuum. The shank of the needle is held at its top end in a piston which slides up and down the dashpot in response to the degree of manifold vacuum. This is directly controlled by the position of the throttle.

With the throttle fully open, the full effect of inlet manifold vacuum is felt by the piston which has an air bleed into the choke tube on the outside of the throttle This causes the piston to rise fully, bringing the needle with it. With the accelerator partially closed only slight inlet manifold vacuum is felt by the piston (although, of course, on the engine side of the throttle the vacuum is now greater), and the piston only rises a little, blocking most of the jet orifice with the metering needle.

To prevent the piston fluttering, and to give a richer mixture when the accelerator is suddenly depressed, an oil damper and light spring are fitted inside the dashpot.

The only portion of the piston assembly to come into contact with the piston chamber or dashpot is the actual central piston rod. All the other parts of the piston assembly, including the lower choke portion, have sufficient clearances to prevent any direct metal to metal contact which is essential if the carburetter is to work properly.

The correct level of the petrol in the carburetter is determined by the level of the float in the float chamber. When the level is correct the float rises and by means of a lever resting on top of it closes the needle valve in the cover of the float chamber. This closes off the supply of fuel from the pump. When the level in the float chamber drops as fuel is used in the carburetter the float sinks. As it does, the float needle comes away from its seat so allowing more fuel to enter the float chamber and restore the correct level.

11. S.U. CARBURETTERS - REMOVAL & REPLACEMENT

1. Referring to Fig. 3.5. take off the air cleaners and free the breather pipe (where fitted).
2. Disconnect the choke cable (3) from the front carburetter by loosening the nut which clamps the inner cable (13) in place on the short actuating arm.
3. Unhook the throttle return springs (4), (the front return spring is shown in the photo), and then disconnect the accelerator control rod (7).

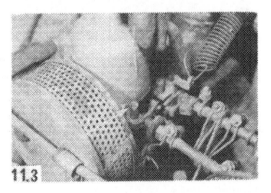

4. Carefully pull off the rubber tubes which carry the fuel (8, 15 and 9) to disconnect the petrol feed pipes.

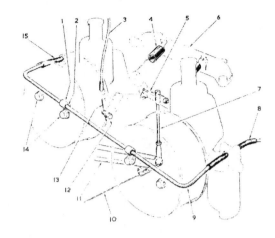

Fig. 3.5. Layout of the twin S.U. carburetters and controls.

5. Undo the nuts which hold the inlet manifold and lift off the twin carburetters complete with their linkages.
6. To replace the carburetters reverse the above procedure using new gaskets throughout. Make sure that with the throttles closed a gap of 0.015 in. exists between the pins on the end of each of the short levers attached to the carburetter interconnecting rod, and the forked lever attached to each of the throttle spindles. (See Fig. 3.6.). The clearance can be gained by loosening the nut and bolt which secures the short lever to the shaft and then turning the short lever on the shaft until the lever pin rests lightly on a 0.015 in. feeler gauge blade fitted between the short lever and the lower arm of the forked lever. Tighten the securing nut and bolt and recheck the clearance.

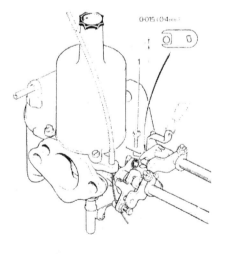

Fig 3.6 Checking the clearance of the forked lever attached to the throttle spindle

FUEL SYSTEM AND CARBURATION

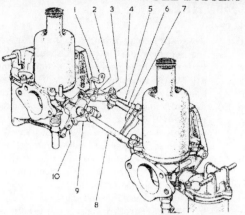

Fig. 3.7. THE S.U. CARBURETTER ACCELERATOR AND CHOKE LINKAGE
1 Bracket for return spring. 2 Stop. 3 Throttle connecting spindle. 4 Short lever. 5 Accelerator lever. 6 Connecting shaft. 7 Lever. 8 Choke (or mixture control shaft. 9 Lever. 10 Choke actuating lever.

12. S.U. CARBURETTERS - DISMANTLING

The S.U. carburetter is a straightforward instrument to service, but at the same time it is a delicate unit and clumsy handling can cause much damage. In particular it is easy to knock the finely tapering needle out of true, and the greatest care should be taken to keep all the parts associated with the dashpot scrupulously clean.

1. Remove the oil dashpot plunger nut from the top of the dashpot.
2. Unscrew the set screws holding the dashpot to the carburetter body, and lift away the dashpot, light spring, and piston and needle assembly.
3. To remove the metering needle from the choke portion of the piston unscrew the sunken retaining screw from the side of the piston choke and pull out the needle. When replacing the needle ensure that the shoulder is flush with the underside of the piston.
4. Release the float chamber from the carburetter by releasing the clamping bolt and sealing washers from the side of the carburetter base.
5. Normally, it is not necessary to dismantle the carburetter further, but if, because of wear or for some other reason, it is wished to remove the jet, this is easily accomplished by removing the clevis pin holding the jet operating lever to the jet head, and then just removing the jet by extracting it from the base of the carburetter. The jet adjusting screw can then be unscrewed together with the jet adjusting screw locking spring.

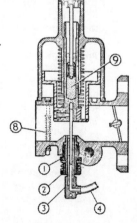

Fig. 3.8. S.U. CARBURETTER
1. Jet gland nut.
2. Jet adjusting nut.
3. Jet assembly and nylon tube
8. Piston lifting pin.
9 Oil well.

6. If the larger locking screw above the jet adjusting screw is removed, then the jet will have to be recentred when the carburetter is reassembled. With the jet screws removed it is a simple matter to release the jet bearing.
7. To remove the throttle and actuating spindle release the two screws holding the throttle in position in the slot in the spindle, slide the throttle out of the spindle and then remove the spindle.
8. Reassembly is a straightforward reversal of the dismantling procedure. It will be necessary to centre the jet. How to do this correctly is described in Section 20.

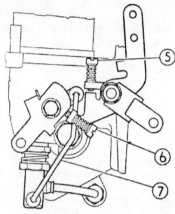

Fig. 3.9. The Jet and Throttle interconnection screws. With the choke right in and the engine idling. Adjust screw (6) to give a clearance of 0.015 in. between the end of the screw and the rocker lever. Always check this when the throttle stop screw (5) is moved.

13. S.U. CARBURETTER FLOAT CHAMBER DISMANTLING, EXAMINATION & REASSEMBLY

1. To dismantle the float chamber, first disconnect the inlet pipe from the fuel pump at the top of the float chamber cover, if this has not already been done.
2. Undo the three screws which hold the float chamber cover in position, and lift off the cover.
3. If it is not wished to remove the float chamber completely and the carburetter is still attached to the engine, carefully insert a thin piece of bent wire under the float and lift the float out.
4. To remove the float chamber from the carburetter body undo the bolt which runs horizontally through the carburetter.
5. Make a careful note of the rubber grommets and washers and on reassembly ensure they are replaced in the correct order. If the float chamber is removed completely it is a simple matter to turn it upside down to drop the float out. Check that the float is not cracked or leaking. If it is, it must be repaired or renewed.
6. The float chamber cover contains the needle valve assembly which regulates the amount of fuel which is fed into the float chamber.
7. One end of the float lever rests on top of the float, rising and falling with it, while the other end pivots on a hinge pin which is held by two lugs. On the float cover side of the float lever is a needle which rises and falls in its brass seating according to the movement of the lever.
8. With the cover in place the hinge pin is held in position by the walls of the float chamber. With the

cover removed the pin is easily pushed out so freeing the float lever and the needle.

9. Examine the tip of the needle and the needle seating for wear. Wear is present when there is a discernible ridge in the chamfer of the needle. If this is evident then the needle and seating must be renewed. This is a simple operation and the hexagon head of the needle housing is easily screwed out.

10. Never renew either the needle or the seating without renewing the other part as otherwise it will not be possible to get a fuel tight joint.

11. Clean the fuel chamber out thoroughly. Reassembly is a reversal of the dismantling procedure detailed above. Before replacing the float chamber cover, check that fuel level setting is correct (see Section 14).

14. S.U. CARBURETTER FLOAT CHAMBER FUEL LEVEL ADJUSTMENT

1. It is essential that the fuel level in the float chamber is always correct as otherwise excessive fuel consumption may occur. On reassembly of the float chamber check the fuel level before replacing the float chamber cover, in the following manner:-

2. Invert the float chamber so that the needle valve is closed. It should now be just possible to place an $\frac{1}{8}$ in. (3.175 mm.) diameter bar between the machined float chamber lip parallel to the float lever hinge, so the face of the float lever just rests on the bar, when the float needle is held fully on its seating.

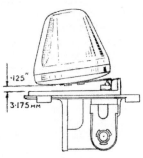

Fig. 3.10. The float level is correct when an $\frac{1}{8}$ in. (3.175 mm.) bar can be slid between the float and the cover flange with the needle valve closed.

3. If the bar lifts the lever or if the lever stands proud of the bar then it is necessary to bend the lever at the bifurcation point between the shank and the curved portion until the clearance is correct. Never bend the flat portion of the lever.

15. S.U. CARBURETTERS - EXAMINATION & REPAIR

The S.U. carburetter generally speaking is most reliable, but even so it may develop one of several faults which may not be readily apparent unless a careful inspection is carried out. The common faults the carburetter is prone to are:-
1. Piston sticking.
2. Float needle sticking.
3. Float chamber flooding.
4. Water and dirt in the carburetter.

In addition the following parts are susceptible to wear after long mileages and as they vitally affect the economy of the engine should be checked and renewed, where necessary, every 24,000 miles.

a) The Carburetter Needle. If this has been incorrectly assembled at some time so that it is not centrally located in the jet orifice, then the metering needle will have a tiny ridge worn on it. If a ridge can be seen then the needle must be renewed. S.U. carburetter needles are made to very fine tolerances and should a ridge be apparent no attempt should be made to rub the needle down with fine emery paper. If it is wished to clean the needle it can be polished lightly with metal polish.

b) The Carburetter Jet. If the needle is worn it is likely that the rim of the jet will be damaged where the needle has been striking it. It should be renewed as otherwise fuel consumption will suffer. The jet can also be badly worn or ridged on the outside from where it has been sliding up and down between the jet bearings everytime the choke has been pulled out. Removal and renewal is the only answer here as well.

c) Check the edges of the throttle and the choke tube for wear. Renew if worn.

d) The washers fitted to the base of the jet, to the float chamber, and to the petrol inlet union may all leak after a time and can cause much fuel wastage. It is wisest to renew them automatically when the carburetter is stripped down.

e) After high mileages the float chamber needle and seat are bound to be ridged. They are not an expensive item to replace and should be renewed as a set. They should never be renewed separately.

16. S.U. CARBURETTERS - PISTON STICKING

1. The hardened piston rod which slides in the centre guide tube in the middle of the dashpot is the only part of the piston assembly (which comprises the jet needle, suction disc, and piston choke) that should make contact with the dashpot.

2. The piston rim and the choke periphery are machined to very fine tolerances so that they will not touch the dashpot or the choke tube walls.

3. After high mileages wear in the centre guide tube (especially on semi-downdraught S.U.s) may allow the piston to touch the dashpot wall. This condition is known as sticking.

4. If piston sticking is suspected or it is wished to test for this condition, rotate the piston about the centre guide tube at the same time sliding it up and down inside the dashpot.

5. If any portion of the piston makes contact with the dashpot wall then that portion of the wall must be polished with metal polish until clearance exists. In extreme cases, fine emery cloth can be used.

6. The greatest care should be taken to remove only the minimum amount of metal to provide the clearance, as too large a gap will cause air leakage and will upset the functioning of the carburetter.

7. Clean down the walls of the dashpot and the piston rim and ensure that there is no oil on them. A trace of oil may be judiciously applied to the piston rod.

8. If the piston is sticking under no circumstances try to clear it by trying to alter the tension of the light return spring.

FUEL SYSTEM AND CARBURATION

17. **S.U. CARBURETTERS - FLOAT NEEDLE STICKING**

    1. If the float needle sticks the carburetter will soon run dry and the engine will stop despite there being fuel in the tank.
    2. The easiest way to check a suspected sticking float needle is to remove the inlet pipe at the carburetter, and spin the engine.
    3. If fuel spurts from the end of the pipe (direct it towards the ground or into a wad of cloth or jar), then the fault is almost certain to be a sticking float needle.
    4. Remove the float chamber and dismantle the valve as detailed on page 69 and clean the housing and float chamber out thoroughly.

18. **S.U. CARBURETTERS - FLOAT CHAMBER FLOODING**

    If fuel emerges from the small breather hole in the cover of the float chamber this condition is known as flooding. It is caused by the float chamber needle not seating properly in its housing; normally because a piece of dirt or foreign matter has become jammed between the needle and the needle housing. Alternatively the float may have developed a leak or be maladjusted so that it is holding open the float chamber needle valve even though the chamber is full of petrol. Remove the float chamber cover, clean the needle assembly, check the setting of the float, and shake the float to verify if any petrol has leaked into it.

19. **S.U. CARBURETTERS - WATER & DIRT IN CARBURETTER**

    1. Because of the size of the jet orifice, water or dirt in the carburetter is normally easily cleared.
    2. If dirt in the carburetter is suspected lift the piston assembly and flood the float chamber. The normal level of fuel should be about $1/16$ in. below the top of the jet and on flooding the carburetter the fuel should well up out of the jet hole.
    3. If very little or no petrol appears, start the engine (the jet is never completely blocked) and with the throttle fully open, blank off the air intake. This will create a partial vacuum in the choke tube and help to suck out any foreign matter from the jet tube. Release the throttle as soon as the engine starts to race. Repeat this procedure several times, stop the engine, and then check the carburetter as detailed in the first paragraph.
    4. If this has failed to do the trick then there is no alternative but to remove and blow out the jet.

20. **S.U. H1 CARBURETTER JET CENTRING**

    The carburetter metering needle is used as a pilot for centring the jet. The piston should therefore be in position, with the dashpot in place, before the jet is centred.

    On the H1 carburetter with the adjusting screw spring removed, screw the adjusting hexagon up into its highest position. The adjusting hexagon is the lower of the two nuts.

    Slide the jet up into position until it is as high as it will go with the base of the jet head against the underside of the adjusting screw.

    The larger hexagon jet screw located above the adjusting nut should now be unscrewed two turns, when it will be found that by gripping the base of the jet and jet adjusting screw that the jet can be moved very slightly laterally.

    Move the jet until the carburetter needle will enter into it fully, and can also be lifted up and will fall under its own weight with a soft click, so entering fully into the jet without any trace of fouling. The jet is now centralised.

    Tighten the large jet screw slightly and check that the needle is still quite free to slide in the jet orifice without binding. Tighten down the jet screw firmly and recheck that the jet is still centralised.

    Remove the jet and the adjusting screw, replace the set screw retaining spring, and return the jet screw and the jet to their positions. It will now be necessary to adjust the carburetter for correct mixture strength.

21. **S.U. CARBURETTERS - ADJUSTMENT & TUNING**

    1. To adjust and tune the S.U. carburetter proceed in the following manner:- Check the colour of the exhaust at idling speed with the choke fully in.
    2. If the exhaust tends to be black, and the tailpipe interior is also black it is a fair indication that the mixture is too rich.
    3. If the exhaust is colourless and the deposit in the exhaust pipe is a very light grey it is likely that the mixture is too weak.
    4. This condition may also be accompanied by intermittent misfiring, while too rich a mixture will be associated with 'hunting'. Ideally the exhaust should be colourless with a medium grey pipe deposit.
    5. Once the engine has reached its normal operating temperature, disconnect the carburetters so each can be worked independently by slackening the bolts on the interconnecting shaft.
    6. Only two adjustments are provided on the S.U. carburetter. Idling speed is governed by the throttle adjusting screw, and the mixture strength by the jet adjusting screw. The S.U. carburetter is correctly adjusted for the whole of its engine revolution range when the idling mixture strength is correct.
    7. Idling speed adjustment is effected by the idling adjusting screw. To adjust the mixture set the engine to run at about 1,000 r.p.m. by screwing in the idling screw. Repeat this procedure for each instrument in turn.
    8. Check the mixture strength by lifting the piston of the carburetter approximately $1/32$ in. (8 mm.) with a thin wire spoke or small screwdriver so as to disturb the airflow as little as possible, when if:
    a) the speed of the engine increases appreciably the mixture is too rich.
    b) the engine speed immediately decreases the mixture is too weak.
    c) the engine speed increases very slightly the mixture is correct.
    9. To enrich the mixture rotate the adjusting screw, which is the screw at the bottom of the carburetter, in an anti-clockwise direction, i.e. downwards. To weaken the mixture rotate the jet adjusting screw in a clockwise direction, i.e. upwards. Only turn the adjusting screw a flat at a time and check the mixture strength between each turn. It is likely that there will be a slight increase or decrease in r.p.m.

after the mixture adjustment has been made so the throttle idling adjusting screw should now be turned so that the engine idles at between 600 and 700 r.p.m.

22. SYNCHRONISATION OF TWIN S.U. CARBURETTERS

1. First ensure that the mixture is correct in each instrument. With twin S.U. carburetters, in addition to the mixture strength being correct for each instrument, the idling suction must be equal on both. It is best to use a vacuum synchronising device such as the Motor Meter synchro-tester. If this is not available, it is possible to obtain fairly accurate synchronisation by listening to the hiss made by the air flow into the intake throats of each carburetter.

2. The aim is to adjust the throttle butterfly disc so that an equal amount of air enters each carburetter. Loosen the clamping bolts on the throttle spindle connections. Listen to the hiss from each carburetter and if a difference in intensity is noticed between them, then unscrew the throttle adjusting screw on the other carburetter until the hiss from both the carburetters are the same.

3. With a vacuum synchronisation device all that it is necessary to do is to place the instrument over the mouth of each carburetter in turn and adjust the adjusting screws until the reading on the gauge is identical for both carburetters.

4. Tighten the clamping bolts on the throttle spindle connections which connect the throttle disc of the two carburetters together, at the same time holding down the throttle adjusting screws against their idling stops. Synchronisation of the two carburetters is now complete.

23. SOLEX B28 21C-2 CARBURETTER - DESCRIPTION

The carburetter fitted to early 948 c.c. models is of the single venturi downdraught type and is sealed so that air for all the jets and the float chamber has to come through the air cleaner. This helps ensure a balanced mixture even if the air cleaner becomes blocked.

A main jet, air correction jet, idling jet, starter air jet, and starter petrol jet, together with the main body, starter valve body, and float chamber comprise the principal components (See Fig. 3.11.) (15).

When starting the engine from cold with the choke out, do not depress the accelerator pedal. Fuel is drawn from the starter jet (23), and emulsified with air, then travels into the inlet manifold.

As soon as the engine fires the starter valve (25) is lifted by suction against its spring allowing in more air. As the engine warms up the choke is pushed in so cutting out the starter unit.

With the engine warm and turning over at idling speed fuel from the pilot jet (15) is emulsified from the pilot jet air bleed. A drilling carries the emulsified fuel past the volume control screw (45) into the airstream just below the throttle (35) which is held slightly open by the throttle adjusting screw (40).

Further opening of the throttle allows manifold depression to draw from a small drilling just above the normal position of the throttle disc when closed,

a further supply of emulsified fuel, this drilling being connected to the pilot jet. As the engine speeds up and the throttle is opened further, depression in the choke tube draws emulsified fuel through the spraying orifice in the centre of the carburetter, the fuel being supplied through the main jet (15), and emulsified by air entering the air correction jet (11) and emulsion tube (12).

With the engine on full throttle the large hole in the bottom of the emulsion tube is exposed and this keeps in balance the ratio of air to fuel from the main jet.

The float in the float chamber rises and falls very slightly according to the amount of petrol being consumed. As soon as the float falls the needle valve opens allowing in more fuel, and when it rises it closes the valve so cutting off the supply.

24. SOLEX B28 21C-2 CARBURETTER - REMOVAL & REPLACEMENT

1. Take off the air cleaner assembly as described in Section 2.

2. Disconnect the choke and throttle cables at the carburetter and undo the vacuum advance and retard pipe union (21) from the carburetter body. NOTE and place on one side the small olive (20).

3. Undo the two nuts and spring washers holding the carburetter in place and lift off. During replacement which is a straightforward reversal of the removal sequence always fit new gaskets.

25. SOLEX B28 21C-2 CARBURETTER - DISMANTLING & REASSEMBLY

1. With time fine sediment deposits will accumulate in the float chamber, drillings and jets, giving rise to poor performance and possible heavy fuel consumption. At intervals of 24,000 miles it is beneficial to clean out the carburetter, and it should be noted that it is not necessary to remove the carburetter from the car in order to gain access to the main jet assembly (16,18), pilot jet (15), and starter jet (23), all of which must be removed for cleaning.

2. With the air cleaner removed and fuel pipe disconnected undo the bolts which hold the float chamber or main body cover (4) in place, and lift it off together with the gasket (5).

3. Take out of the float chamber (19) the float (10), toggle (9), and pin (8) and check the float for leaks. Also check the functioning of the needle valve (7) and renew it if worn.

4. Take out all the jets (18,16,15 and 23) and blow all dirt out of the carburetter using compressed air. On no account try poking wire through any of the jets or drillings. If need be the jets can be poked with a fine nylon bristle.

5. Replacement is a straightforward reversal of the removal sequence. Always use new gaskets and ensure the same number of fibre washers, as were removed, are replaced between the top cover and the needle valve.

26. SOLEX B28 21C-2 - CARBURETTER ADJUSTMENT

1. Apart from fitting jets of different sizes only two adjustments are available. The idling mixture strength can be reduced by screwing in the volume

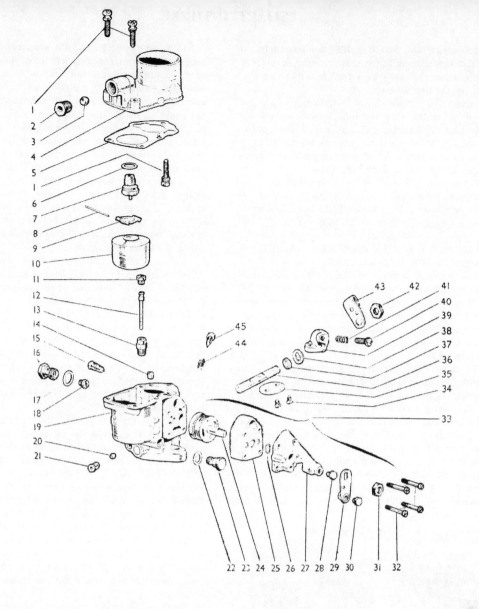

Fig. 3.11. EXPLODED VIEW OF THE SOLEX B28ZIC-2 CARBURETTER

1 Screws—float chamber cover top to main body. 2 Fuel union nut. 3 Fuel union olive. 4 Float chamber cover. 5 Gasket. 6 Needle valve washer. 7 Needle valve. 8 Float toggle pin. 9 Float toggle. 10 Float. 11 Sir correction jet. 12 Emulsion tube. 13 Head of emulsion tube. 14 Pilot jet air bleed. 15 Pilot jet. 16 Main jet holder. 17 Washer. 18 Main jet. 19 Float chamber main body. 20 Ignition union olive. 21 Ignition union. 22 Washer. 23 Starter fuel jet. 24 Starter valve assembly. 25 Starter valve body. 26 Circlip. 27 Starter cover. 28 Cable locking screw. 29 Starter lever. 30 Locking screw. 31 Nut. 32 Starter unit fixing screws. 33 Starter assembly. 34 Throttle fixing screws. 35 Throttle. 36 Throttle spindle. 37 Sealing washer. 38 Distance washer. 39 Throttle spindle abutment plate. 40 Slow running adjustment screw. 41 Spring. 42 Nut. 43 Throttle lever. 44 Spring 45 Volume control screw.

control screw (45), and the idling speed can be increased by screwing in the slow running screw (40). In both cases the opposite effect is achieved by screwing the screws out.

2. Adjust the carburetter with the engine hot. Set the engine by the slow running screw (40) to run a little faster than normal, and screw the volume control screw out until the engine begins to hunt. (Mixture over-rich). Slowly screw the volume control in until the hunting just stops.

3. If the engine speed is too fast reset the slow running screw (550 r.p.m. is ideal), and if the engine now hunts a little screw in the volume control screw a fraction further till all is correct.

27. SOLEX B30 PSE1 CARBURETTER - REMOVAL, DISMANTLING & REPLACEMENT

1. The carburetter fitted to all models of the Herald 1200, and 12/50 is very similar to the Solex B28-21C-2 instrument especially in operating principles and is of the single venturi downdraught type incorporating an accelerator pump and economy device and a choke valve of the semi-automatic strangler type.

2. To remove the carburetter from the engine first remove the air cleaner, slacken back the clamp securing the cleaner hose to the carburetter top and lift off. Referring to Fig. 3.12. disconnect the fuel feed pipe union (72) and the distributor vacuum pipe (71) from the carburetter, and then disconnect the choke control cable (68) at the operating cam (40), detaching the clip (37) securing the outer cable in position. Free the throttle cable by loosening the screw (66). Remove the two nuts and spring washers (70) which hold the flange to the manifold and lift off the carburetter.

Fig. 3.12. External view of the Solex B30 PSE carburetter. 13 Slow running adjustment screw. 14 Slow running jet. 22 Volume control screw. 32 Circlip. 37 Abutment bracket. 40 Pinch screw. 41 Pinch screw. 65 Accelerator connection. 66 Screw. 67 Bracket. 68 Choke cable. 69 Accelerator cable. 70 Securing nut. 71 Rubber sleeve (advance and retard pipe). 72 Fuel inlet union.

3. To dismantle the carburetter first unscrew the five screws with spring washers securing the float chamber cover to the body, and lift away the body together with the gasket. Lift out the float arm and hinge pin and remove the float. Detach the split pin which retains the pushrod and spring, and then remove the four screws holding the accelerator pump in position, lifting away the pump body and operating arm together with the diaphragm and return spring.

4. Remove the spring clip securing the throttle link, unscrew the screw securing the link to the choke operating cam, and lift the link clear. Remove the bolt securing the operating cam and return spring and unscrew the cheese-headed screw holding the choke cable abutment bracket.

5. Unscrew the idling jet, high speed air correcting jet and emulsion tube assembly and lift off the accelerator pump discharge nozzle, then unscrew the anti-siphon valve and lift away the ball.

6. It is not normally necessary to remove the discharge beak but if this is removed then it will be essential to lock the taper end screw with lead shot when it is replaced. The shot should be inserted through the vertical drilling which will be located above the taper end screw.

7. Unscrew the bolt and flat washer which will give access to the main jet and allow this to be removed, and then remove the throttle plate or spindle, extract the two screws which hold this plate in position and withdraw the spindle and plate.

8. First refit the throttle spindle into the carburetter body and then fit the throttle plate ensuring that the mark (') stamped on the plate is away from the accelerator pump and facing downwards when the throttle is closed. Fit and tighten the two securing screws and then lightly peen over the ends to prevent any possible loosening.

9. Replace the main jet and the blanking plug with its washer followed by the idling jet, main air correction jet and emulsion tube assembly and the accelerator pump discharge nozzle. Replace the ball and refit the anti-siphon valve, and then refit the choke cable abutment bracket, tightening the cheese-headed retaining screw finger-tight, while locating the choke operating cam and its retracting spring in position by using the hexagon headed bolt as a guide. Ensure that the inner end of the spring is located in the slot in the abutment bracket and that the outer end is against the 'V' in the operating cam. Both screws should then be tightened securely.

10. Replace the choke link, securing it to the choke-throttle link with a spring clip and to the choke operating cam with the adjustment bolt. Refit the pushrod to the spindle link, and secure with a spring clip. Replace the pushrod spring, attach the accelerator pump operating lever to the pushrod and secure with the split pin.

11. Install the return spring and diaphragm in the pump housing, refit the assembly to the carburetter body with the four retaining screws and then check the action of the return spring.

12. Drop the float into the float chamber with the cup washer to the top. Fit the float lever and hinge pin in position with the curve on the end of the lever towards the float.

13. Fit a new gasket on the top of the float chamber and offer to the cover while holding the choke plate in the fully open position. Secure firmly with the five screws and spring washers.

14. Once the carburetter is assembled it is fitted to the engine by first locating a new gasket on the manifold flange and then bolting down the carburetter flange, making quite certain that it is flat and level on the stud before tightening the nuts.

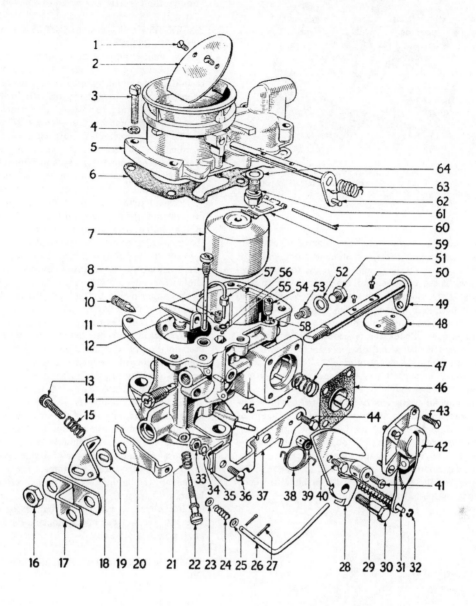

Fig. 3.13. EXPLODED VIEW OF THE SOLEX B30 PSE1 CARBURETTER

1 Screw. 2 Strangler. 3 Screw. 4 Spring washer. 5 Top cover. 6 Gasket. 7 Float. 8 Air correction jet. 9 Econostat fuel jet. 10 Spraying bridge retaining screw. 11 Body assembly. 12 Spraying bridge. 13 Slow running adjustment screw. 14 Slow running fuel jet. 15 Spring. 16 Nut. 17 Throttle lever. 18 Stop lever. 19 Slotted washer. 20 Strangler—inter-connection lever. 21 Spring. 22 Volume control screw. 23 Washer. 24 Spring. 25 Washer. 26 Strangler inter-connection pushrod. 27 Split pin. 28 Strangler operating cam. 29 Spring. 30 Pivot bolt. 31 Accelerator pump pushrod. 32 Circlip. 33 Nut. 34 Spring washer. 35 Cable clip. 36 Screw. 37 Abutment bracket. 38 Spring. 39 Solderless nipple. 40 Pinch screw. 41 Pinch screw. 42 Pump cover and lever assembly. 43 Screw. 44 Set screw. 45 Non-return ball valve. 46 Pump diaphragm. 47 Diaphragm spring. 48 Throttle butterfly. 49 Throttle spindle. 50 Screw. 51 Main jet access. 52 Fibre washer. 53 Main jet. 54 Pump chamber non-return valve body. 55 Non-return ball valve. 56 Fibre washer. 57 Accelerator pump jet. 58 Pump chamber non-return valve. 59 Float lever. 60 Float lever pivot. 61 Needle valve. 62 Strangler cam follower and spindle. 63 Return spring. 64 Fibre washer.

CHAPTER THREE

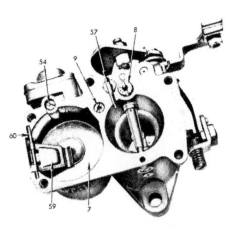

Fig. 3.14. The Solex B30 PSE 1 carburetter with the top cover removed showing: 7 Float. 8 Air correction jet. 9 Ecoustat fuel jet. 54 Valve body. 57 Accelerator pump jet. 59 Float arm. 60 Hinge pin.

15. Connect up the distributor vacuum pipe to the rubber connection on the right-hand side, and refit the fuel pipe from the pump. Refit the throttle control rod to the upper end of the throttle lever connecting rod. Connect the choke control cable and tighten the clamp. Pass the cable inner wire through the choke operating cam trunnion and tighten the clamp screw. Check the operation of the choke to ensure that it opens and closes correctly and that there is a limited amount of slack in the cable when the control is pushed right home. Refit the air cleaner to the carburetter.

16. The choke control cable is adjusted at the operating cam to provide at least $1/8$ in. free movement in the cable when the control is pushed right home.

17. The correct degree of throttle opening when the choke plate is closed for starting is obtained by placing a number 71 drill (.027 in. between the edge of the throttle plate and the carburetter body at right angles to the throttle spindle.

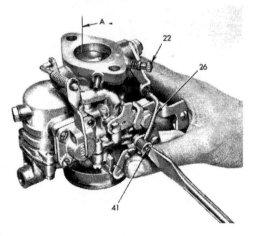

Fig. 3.15. To ensure the throttle opens very slightly when the choke is pulled out place a piece of 0.027 in. (0.7 mm.) wire at 'A' and adjust the rod (26) so the throttle is held in this position when the rod is removed.

18. Alternatively this setting can be obtained by screwing the throttle stop screw in three turns from the position which it abuts the throttle plate stop when the throttle is closed. It will be necessary to remove the throttle screw and remove the spring.

28. SOLEX B30 PSE1 CARBURETTER - SLOW RUNNING ADJUSTMENT

1. To obtain the best slow-running adjustment the engine should be tuned against a vacuum gauge connected to the inlet manifold. To enable this to be accomplished a blanking plug will be found in the carburetter flange, and it is first necessary to remove this plug and fit in a screwed adaptor.

2. To this is connected a plastic tube which in turn is connected to the rear of a vacuum gauge.

3. Before commencing any adjustments make sure that the air cleaner is clean and not blocked, since this will affect the amount of air in the fuel mixture.

4. It is important to appreciate that the correct use of a vacuum gauge allows for checking the condition of the engine, the correct timing of the ignition and the general running characteristics as well as the correct tuning of the carburetter, and these points are dealt with in the Section dealing with tuning the engine for satisfactory running.

5. No adjustments to the carburetter should be attempted until the engine has been run to bring it up to working temperature.

6. To obtain correct mixture, screw in the throttle stop until a reasonable idling speed is attained, and then turn the volume control screw either one way or the other until the maximum vacuum reading, usually between 19° and 22° is attained, and the needle remains steady. Ease back the throttle screw and continue adjustment of the volume control screw, watching the vacuum gauge needle at each adjustment until the maximum reading is obtained with the engine at a reasonable slow-running speed.

7. To attain the most satisfactory result it may be necessary to adjust the ignition setting described later in Chapter 4.

8. If a satisfactory reading cannot be obtained the reasons are set out in the section dealing with the engine fault diagnosis.

29. STROMBERG 150 CD CARBURETTER - DESCRIPTION

All Herald 13/60 models are fitted with a single Stromberg 150 CD (constant depression) carburetter which is fitted with a horizontal single choke. The carburetter works in similar fashion to the S.U. instrument but there are important differences. The main two are that the float chamber is attached to and is directly under the main body of the carburetter, and instead of a piston which rises and falls as in the S.U., a flexible diaphragm raises and lowers the needle. (See Fig. 3.16.)

When starting from cold pulling the choke out rotates the lever (71) and bar (28). (All references are to Fig. 3.17.) which lifts the piston type air valve and needle (9, 11). On no account should the accelerator be pressed before the engine fires. The metering needle tapers slightly and fits into the jet orifice (80). The higher the needle is raised the richer the mixture becomes, because of the in-

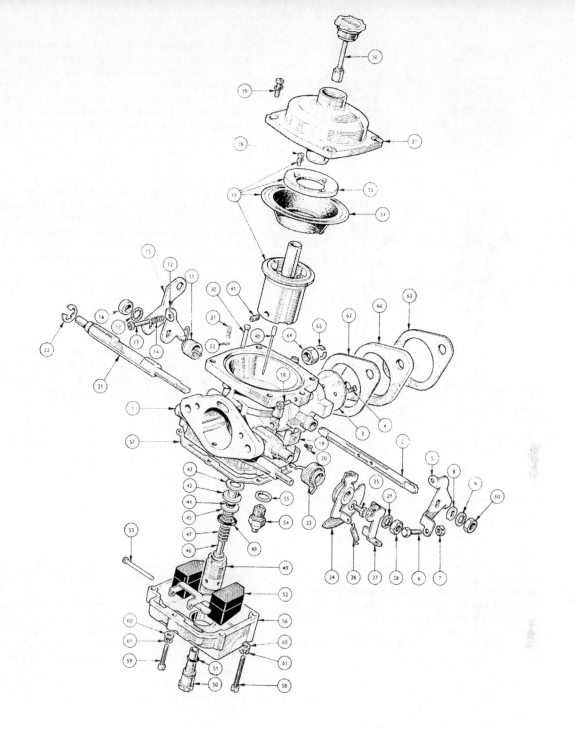

Fig. 3.16. EXPLODED VIEW OF THE ZENITH-STROMBERG 150CD CARBURETTER
1 Main body assembly. 2 Throttle spindle. 3 Throttle. 4 Screw—throttle to spindle. 5 Lever—throttle stop and fast idle. 6 Fast idle adjustment screw. 7 Nut. 8 Washer. 9 Shakeproof washer. 10 Nut. 11 Throttle return spring. 12 Throttle stop. 13 Throttle stop screw. 14 Spring. 15 Throttle lever. 16 Nut. 17 Shakeproof washer. 18 Screw. 19 Main body ventilation hole rap. 20 Screw. 21 Starter bar. 22 Retaining ring. 23 Spring. 24 Lever. 25 Screw. 26 Return spring. 27 Choke lever. 28 Nut. 29 Shakeproof washer. 30 Pin. 31 Spring. 32 Clip. 33 Piston and piston diaphragm assembly. 34 Diaphragm. 35 Washer. 36 Screw. 37 Suction chamber cover. 38 Damper assembly. 39 Screw. 40 Metering needle. 41 Locking screw. 42 Jet orifice bushing. 43 Bushing washer. 44 'O' ring. 45 Washer. 46 Jet. 47 Spring. 48 'O' ring. 49 Screw. 50 Adjusting screw. 51 'O' ring. 52 Nylon float and arm. 53 Pin. 54 Needle. 55 Washer. 56 Float chamber. 57 Float chamber gasket. 58 Screw (long)—float chamber to body. 59 Screw (short)—float chamber to body. 60 Washer. 61 Spring washer. 62 Washer. 63 Washer. 64 Nut. 65 Lock washer.

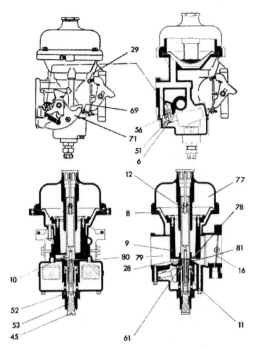

Fig. 3.17. SECTION VIEWS OF STROMBERG 150 CD CARBURETTER

6 Needle. 8 Diaphragm. 9 Air valve. 10 Locking screw. 11 Needle. 12 Damper. 16 Throttle. 28 Starter bar. 29 Screw. 45 Adjusting screw. 51 Float assembly. 52 'O' ring. 53 Bushing screw. 61 Bushing. 69 Screw. 71 Lever. 77 Chamber. 78 Air valve drilling. 79 Bore. 80 Jet orifice. 81 Bridge.

creased discharge area available at the mouth of the jet.

At the same time as the choke is pulled out a cam on the lever (71) opens the throttle beyond its normal idling position, the extent to which the throttle is opened, depending on the setting of the fast idle screw (69).

As soon as the engine starts, increased vacuum in the inlet manifold lifts the piston (9), so weakening the mixture to prevent the engine stalling because of over-richness. As the engine warms up the choke is pushed in gradually until the lever (71) is back in its normal position.

When the throttle is opened the decrease in pressure in the induction manifold is also felt in the suction chamber (77) above the piston (9) because of the drilling (78) in the piston base. Because the suction chamber is sealed from the main body (78) by means of a diaphragm (8) pressure above the piston is decreased and the piston lifts. As this happens, the choke area is increased and the depression above the piston is reduced. In this way the pressure drop across the jet orifice and air velocity remains virtually constant at all speeds irrespective of throttle opening. Thus the piston rises and falls, and the choke area varies as the engine's demands alter. The tapered needle protruding from the base of the piston moves up and down in the jet, and the space between the needle and the jet increases and decreases as the piston rises and falls.

This of course, varies the mixture strength, and accounts for the influence of needle shape on engine performance.

Snap acceleration requires a richer mixture. The Stromberg is now faced with an additional difficulty. Sudden throttle opening could increase suction over the piston to such an extent that the piston would rise very rapidly to the top of the chamber. This would lower the depression in the jet and cause a weakening of the mixture just when it should be enriched. In fact, this difficulty is avoided by the inclusion of a piston damper in the design. The oil well in the middle of the piston rod has a spindle in it which is attached to the screw cap. On the end of the spindle is a sleeve which provides opposition to the oil flow as the piston rises, but not when it falls. This obstruction decreases the speed of the piston's rise and thus increases mixture strength when the throttle is opened suddenly. The level of this S.A.E. 20 oil should be up to $1/4$ in. of the end of the well.

30. STROMBERG 150 CD CARBURETTER - ADJUSTMENTS

1. As there is no separate idling jet, the mixture for all conditions is supplied by the main jet and variable choke. Thus the strength of the mixture throughout the range depends on the height of the jet in the carburetter body and when the idling mixture is correct, the mixture will be correct throughout the range. A slotted nut (45) at the base of the carburetter increases or decreases the strength of the mixture. Turning the nut clockwise raises the jet and weakens the mixture. Turning the jet anti-clockwise enriches the mixture. The idling speed is controlled by the throttle stop screw (29).

2. To adjust a Stromberg 150 CD carburetter from scratch, run the engine until it is at its normal working temperature and then remove the air cleaner and damper and cap assembly. Press and hold the piston down with a length of wire held in the oil well so the underside of the piston rest on the bridge of the choke (81). With a coin screw up clockwise the slotted jet adjustment nut (45) until the head of the jet can be felt to just touch the underside of the piston.

3. Then, from the position referred to in the previous paragraph turn the jet screw anti-clockwise three full turns. This will give an approximate setting. Start the engine and adjust the idle stop screw (29) so the engine runs fairly slowly and smoothly (about 600/650 r.p.m.) without rocking on its mountings. To get the engine to run smoothly at this speed it may be necessary to turn the jet adjuster nut a small amount in either direction.

4. To test if the correct setting has been found lift the piston $1/32$ in. through the air intake with an electrical spanner or wire spoke. This is a very small amount and care should be taken to lift the piston only fractionally. If the engine speed rises the mixture is too rich and if it hesitates or stalls it is too weak. Re-adjust the jet adjusting nut and re-check. All is correct when the engine speed does not increase when the piston is lifted the requisite amount.

31. STROMBERG 150 CD CARBURETTER - FLOAT CHAMBER FUEL LEVEL ADJUSTMENT

1. Take off the air cleaner and then remove the carburetter from the engine.

FUEL SYSTEM AND CARBURATION

2. Undo the five small screws which hold the float chamber to the base of the carburetter body.

3. Turn the carburetter body upside down and accurately measure the highest point of the floats which should be 18 mm. above the flange normally adjacent to the float chamber (see Fig. 3.18.). During this operation ensure that the needle is against its seating. To reset the level carefully bend the tag which bears against the end of the needle.

Fig. 3.18. The float level is correct when the distance between the bottom of the float and the flange (the distance between the arrows 'A') is 18 mm.

32. STROMBERG 150 CD CARBURETTER - DISMANTLING & REASSEMBLY

1. Take off the air cleaner, disconnect the choke and accelerator controls at the carburetter, also the vacuum advance and retard pipe and undo the nuts and spring washers holding the carburetter in place.

2. All figures in brackets refer to Fig. 3.16. With the carburetter on the bench undo and remove the clamper cap and plunger (38). Then undo the four screws (39) which hold the suction chamber cover (37) in place and lift off the cover.

3. The piston (33) complete with needle (40) and diaphragm (34) is then lifted out. Handle the assembly with the greatest of care as it is very easy to knock the needle out of true.

4. The bottom of the float chamber (56) is removed by undoing the five screws (58, 59), and the spring and flat washers (60, 61) which hold it in place. Take out the pin (53) and remove the float assembly (52).

5. If wished the needle (40) may be removed from the piston (33) by undoing the grub screw (41).

6. To remove the diaphragm (34) from the piston simply undo the four screws and washers (36) which hold the diaphragm retaining ring (35) in place.

7. The jet (46) and associated parts are removed after the jet locking nut has been undone.

8. On reassembly there are several points which should be noted particularly. The first is that if fitting a new needle to the piston ensure it has the same markings as the old stamped on it, and fit it so that the needle shoulder is perfectly flush with the base of the piston.

9. Thoroughly clean the piston and its cylinder in paraffin and when replacing the jet centralise it as described in paragraphs 10 to 12.

10. Refit the jet and associated parts, lift the piston and tighten the jet assembly.

11. Turn the mixture adjusting nut clockwise until the tip of the jet is just above the choke bridge. Now loosen the jet assembly about one turn so as to free the orifice bush (42).

12. Allow the piston to fall. As it descends the needle will enter the orifice and automatically centralise it. With the needle still in the orifice tighten the jet assembly slowly, frequently raising and dropping the piston $1/4$ in. to ensure the orifice bush has not moved. Finally check that the piston drops freely without hesitation and hits the bridge with a soft metalic sound.

13. Make sure that the holes in the diaphragm line up with the screw holes in the piston and retaining ring, and that the diaphragm is correctly positioned. Reassembly is otherwise a straightforward reversal of the dismantling sequence.

33. FUEL TANK - REMOVAL & REPLACEMENT

1. Remove the filler cap and unscrew the drain plug, which is located in the floor behind the left-hand wheel arch, and drain the contents into a suitable container. When empty replace the plug and washer securely.

2. Take off the rubber cover from the tank sender unit and disconnect the wires from their terminals.

3. Free the rubber tube from the fuel pipe at the front top cover of the tank.

4. Undo and remove the four screws and washers which hold the tank in place in the rear wing, lift the tank a little so the drain plug clears the hole in the floor, and then lift the tank out, at the same time freeing the filler neck from the rubber grommet in the wing.

5. Replacement is quite straightforward and is a reversal of the removal sequence. To avoid water getting into the boot reseal the drain plug boss to the floor with a new sponge rubber grommet.

34. FUEL GAUGE - REMOVAL & REPLACEMENT

1. On models where the fuel gauge forms part of the dial of the speedometer this must be removed before the fuel gauge can be taken off. Pull off the snap connectors to the gauge from behind the fascia, disconnect the speedometer and trip cables, and pull out the dial illuminating bulbs.

2. Undo the knurled nuts, and take off the spring washers, bridge piece and abutment ring and pull out the speedometer. Undo the screws which hold the fuel gauge to the speedometer and remove the gauge.

3. On models with a separate fuel gauge disconnect the snap connectors, undo the two knurled nuts and spring washers which hold the bridge clamp to the gauge, take off the bridge piece and remove the gauge from the fascia.

4. Replacement in both instances is quite straightforward and is a reversal of the removal sequence.

CHAPTER THREE

35. SOLEX CARBURETTERS — LACK OF FUEL AT ENGINE

If it is not possible to start the engine, first positively check that there is fuel in the fuel tank, and then check the ignition system as detailed in Chapter 4. If the fault is not in the ignition system then disconnect the fuel inlet pipe from the carburetter and turn the engine over by the starter relay switch.

If petrol squirts from the end of the inlet pipe, reconnect the pipe and check that the fuel is getting to the float chamber. This is done by unscrewing one of the jets on the float chamber portion of the carburetter, when fuel should run out.

If fuel runs out then it is likely that there is a blockage in the starting jet, which should be removed and cleaned.

No fuel indicates that there is no fuel in the float chamber, and this will be caused either by a blockage in the pipe between the pump and float chamber or a sticking float chamber valve.

If it is decided that it is the float chamber valve that is sticking, remove the bolts that secure the float chamber cover in position, remove the fuel inlet pipe, and lift the cover, complete with valve and floats, away.

Remove the valve spindle and valve and thoroughly wash them in petrol. Petrol gum may be present on the valve or valve spindle and this is usually the cause of a sticking valve. Replace the valve in the needle valve assembly, ensure that it is moving freely, and then reassemble the float chamber. It is important that the same washer be placed under the needle valve assembly as this determines the height of the floats and therefore the level of petrol in the chamber.

Reconnect the fuel pipe and refit the air cleaner.

If no petrol squirts from the end of the pipe leading to the carburetter then disconnect the pipe leading to the inlet side of the fuel pump. If fuel runs out of the pipe then there is a fault in the fuel pump, and the pump should be checked as has already been detailed.

No fuel flowing from the tank when it is known that there is fuel in the tank indicates a blocked pipe line. The line to the tank should be blown out. It is unlikely that the fuel tank vent would become blocked, but this could be a reason for the reluctance of the fuel to flow. To test for this, blow into the tank down the fill orifice. There should be no build up of pressure in the fuel tank, as the excess pressure should be carried away down the vent pipe.

36. SOLEX CARBURETTERS - WEAK MIXTURE

If the fuel/air mixture is weak there are six main clues to this condition:-
1. The engine will be difficult to start and will need much use of the choke, stalling easily if the choke pushed in.
2. The engine will overheat easily.
3. If the sparking plugs are examined (as detailed in Chapter 4), they will have a light grey/white deposit on the insulator nose.
4. The fuel consumption may be light.
5. There will be a noticeable lack of power.
6. During acceleration and on the over-run there will be a certain amount of spitting back through the carburetter.

As the carburetter is of the fixed jet type, these faults are invariably due to circumstances outside the carburetter. The only usual fault likely in the carburetter is that one or more of the jets may be partially blocked. If the car will not start easily but runs well at speed, then it is likely that the starting jet is blocked, whereas if the engine starts easily but will not rev. then it is likely that the main jets are blocked.

If the level of petrol in the float chamber is low this is usually due to a sticking valve or incorrectly set floats.

Air leaks either in the fuel lines, or in the induction system should also be checked for. Also check the distributor vacuum pipe connection as a leak in this is directly felt in the inlet manifold.

The fuel pump may be at fault as has already been detailed.

37. SOLEX CARBURETTERS - RICH MIXTURE

If the fuel/air mixture is rich there are also six main clues to this condition.
1. If the sparking plugs are examined (as detailed in Chapter 4), they will be found to have a black sooty deposit on the insulator nose
2. The fuel consumption will be heavy.
3. The exhaust will give off a heavy black smoke, especially when accelerating.
4. The interior deposits on the exhaust pipe will be dry, black and sooty (if they are wet, black and sooty this indicates worn bores, and much oil being burnt).
5. There will be a noticeable lack of power.
6. There will be a certain amount of backfiring through the exhaust system.

The faults in this case are usually in the carburetter and the most usual is that the level of petrol in the float chamber is too high. This is due either to dirt behind the needle valve, or a leaking float which will not close the valve properly, or a sticking needle.

With a very high mileage (or because someone has tried to clean the jets out with wire), it may be that the jets have become enlarged.

If the air correction jets are restricted in any way the mixture will become very rich.

Occasionally it is found that the choke control is sticking or has been maladjusted.

Again, occasionally the fuel pump pressure may be excessive so forcing the needle valve open slightly until a higher level of petrol is reached in the float chamber.

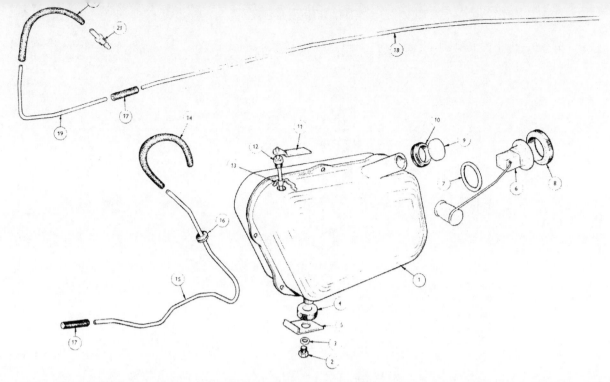

Fig. 3.17. EXPLODED VIEW OF THE PETROL TANK AND ASSOCIATED COMPONENTS FITTED TO THE HERALD SALOON, COUPE & CONVERTIBLE
1 Petrol tank. 2 Petrol drain plug. 3 Drain plug washer. 4 Seal. 5 Drain plug bracket. 6 Gauge. 7 Washer. 8 Gauge unit cover. 9 Filler cap. 10 Grommet. 11 Pipe. 12 Nut. 13 Seal. 14 Connector. 15 Pipe. 16 Grommet. 17 Connector. 18 Pipe. 19 Pipe. 20 Connector. 21 Pipe to fuel pump.

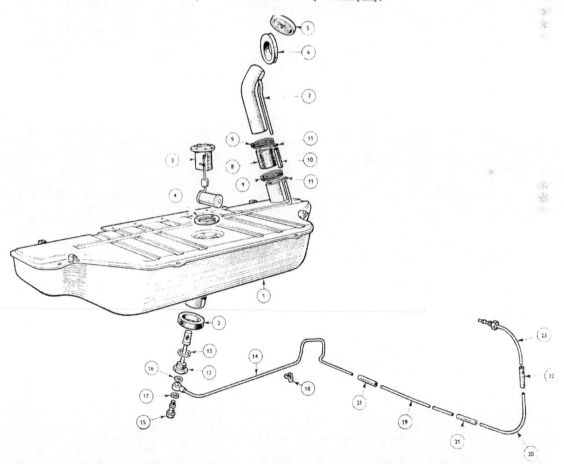

Fig. 3.20. EXPLODED VIEW OF THE PETROL TANK AND ASSOCIATED COMPONENTS FITTED TO THE ESTATE CAR
1 Petrol tank. 2 Sealing ring. 3 Bi-metal gauge unit. 4 Washer. 5 Filler cap. 6 Grommet. 7 Top filler pipe. 8 Connection. 9 Clip. 10 Vent pipe connection. 11 Clip. 12 Pipe. 13 Sealing washer. 14 Pipe assembly. 15 Banjo bolt. 16 Washer. 17 Washer. 18 Clip. 19 Pipe. 20 Pipe. 21 Connector. 22 Front pipe connector. 23 Pipe assembly.

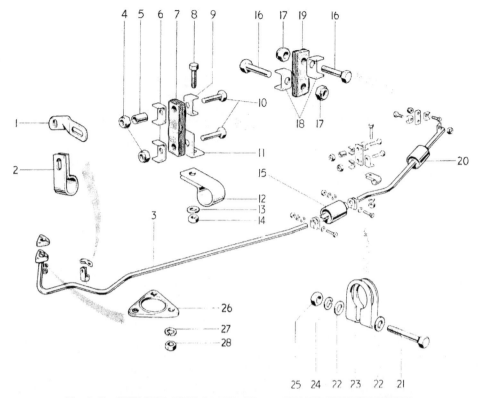

Fig. 3.21. EXPLODED VIEW OF THE 948 c.c. HERALD EXHAUST SYSTEM
1 Bracket—front pipe to clutch housing. 2 Clip. 3 Front exhaust pipe. 4 Nyloc nut. 5 Distance piece. 6 Plate. 7 Fabric strap. 8 Bolt. 9 Plate. 10 Bolts. 11 Angle bracket. 12 Pipe clip. 13 Washer. 14 Nut. 15 Intermediate silencer. 16 Bolt. 17 Nyloc nut. 18 Plate. 19 Fabric strap. 20 Rear silencer and tail pipe. 21 Pinch bolt. 22 Plain washer. 23 Clip. 24 Spring washer. 25 Nut. 26 Flange gasket. 27 Spring washer. 28 Nut.

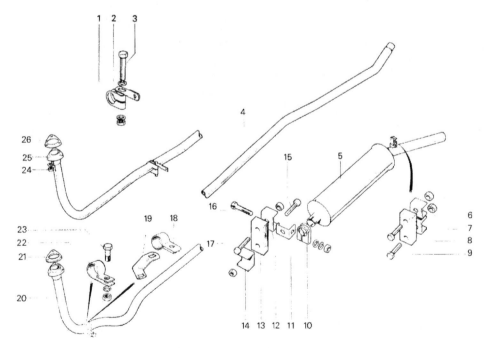

Fig. 3.22. EXPLODED VIEW OF THE EXHAUST SYSTEM FITTED TO 1200, 12/50, 13/60 MODELS
1 Pipe clip. 2 Bracket. 3 Bolt. 4 Exhaust pipe. 5 Exhaust silencer and rear pipe. 6 Clamp plate. 7 Clamp plate. 8 Flexible mounting strip. 9 Setscrew. 10 Clip. 11 Mounting bracket. 12 Clamp plate. 13 Flexible mounting strip. 14 Clamp plate. 15 Setscrew. 16 Bolt. 17 Setscrew. 18 Clip. 19 Bracket. 20 Front exhaust pipe. 21 Gasket. 22 Clip. 23 Bolt. 24 Nut. 25 Front exhaust pipe. 26 Gasket.

FUEL SYSTEM AND CARBURATION
FAULT FINDING CHART

Cause	Trouble	Remedy
SYMPTOM:	FUEL CONSUMPTION EXCESSIVE	
Carburation and ignition faults	Air cleaner choked and dirty giving rich mixture	Remove, clean and replace air cleaner.
	Fuel leaking from carburettor(s), fuel pumps, or fuel lines	Check for and eliminate all fuel leaks. Tighten fuel line union nuts.
	Float chamber flooding	Check and adjust float level.
	Generally worn carburettor(s)	Remove, overhaul and replace.
	Distributor condenser faulty	Remove, and fit new unit.
	Balance weights or vacuum advance mechanism in distributor faulty	Remove, and overhaul distributor.
Incorrect adjustment	Carburettor(s) incorrectly adjusted mixture too rich	Tune and adjust carburettor(s).
	Idling speed too high	Adjust idling speed.
	Contact breaker gap incorrect	Check and reset gap.
	Valve clearances incorrect	Check rocker arm to valve stem clearances and adjust as necessary.
	Incorrectly set sparking plugs	Remove, clean, and regap.
	Tyres under-inflated	Check tyre pressures and inflate if necessary.
	Wrong sparking plugs fitted	Remove and replace with correct units.
	Brakes dragging	Check and adjust brakes.
SYMPTOM:	INSUFFICIENT FUEL DELIVERY OR WEAK MIXTURE DUE TO AIR LEAKS	
Dirt in system	Petrol tank air vent restricted	Remove petrol cap and clean out air vent.
	Partially clogged filters in pump and carburettor(s)	Remove and clean filters.
	Dirt lodged in float chamber needle housing	Remove and clean out float chamber and needle valve assembly.
	Incorrectly seating valves in fuel pump	Remove, dismantle, and clean out fuel pump.
Fuel pump faults	Fuel pump diaphragm leaking or damaged	Remove, and overhaul fuel pump.
	Gasket in fuel pump damaged	Remove, and overhaul fuel pump.
	Fuel pump valves sticking due to petrol gumming	Remove, and thoroughly clean fuel pump.
Air leaks	Too little fuel in fuel tank (Prevalent when climbing steep hills)	Refill fuel tank.
	Union joints on pipe connections loose	Tighten joints and check for air leaks.
	Split in fuel pipe on suction side of fuel pump	Examine, locate, and repair.
	Inlet manifold to block or inlet manifold to carburettor(s) gasket leaking	Test by pouring oil along joints - bubbles indicate leak. Renew gasket as appropriate.

CHAPTER FOUR

IGNITION SYSTEM

CONTENTS

General Description...	1
Contact Breaker Adjustment ...	2
Removing & Replacing Contact Breaker Points ...	3
Condenser Removal, Testing & Replacement...	4
Distributor Lubrication ...	5
Distributor Removal & Replacement ...	6
Distributor Dismantling ...	7
Distributor Inspection & Repair ...	8
Distributor Reassembly ...	9
Ignition Timing ...	10
Sparking Plugs & Leads ...	11
Ignition System Fault Finding...	12
Ignition System Fault Symptoms ...	13
Fault Diagnosis - Engine Fails to Start...	14
Fault Diagnosis - Engine Misfires...	15

SPECIFICATIONS

Sparking Plugs
    948 c.c. engines ... Lodge HLN
    Herald 1200 and 12/50 ... Lodge CLNY
    from engine Nos. GA.185794E
        and GD.65575E ... Champion L87Y
    Herald 13/60... Champion N-9Y
    Plug gap ... .025 in. (.64 mm.)
    Firing order ... 1, 3, 4, 2.

Coil ... Lucas LA12 or HA12
    Resistance at 20°C (68°F) in primary winding   3.1 to 3.5 ohms.
    Consumption - ignition switched on ... 3.9 amps.
    At 1,000 r.p.m. ... 1.25 amps.

Distributors
    948 c.c. engines Low and High C .. ... Lucas DM2
    Herald 1200 and 12/50 High C ... Lucas DM2
                           Low C ... Lucas DM2
    From 1964 ... Lucas 25D4
    Herald 13/60 .. ... Lucas 25D4
    Contact points gap setting ... .014 to .016 in. (.35 to .40 mm.)
    Rotation of rotor ... Anti-clockwise
    Automatic advance.. ... Centrifugal and vacuum
    Contact breaker spring - Tension.. ... 18 to 24 ozs.
    Condenser capacity.. ... .18 to .24 microfarad
    Serial numbers DM2 ...
    948 c.c. 7 to 1 compression ratio ... 40658
    948 c.c. 8 to 1 compression ratio ... 40637
    948 c.c. 7.4 to 1 and 8.5 to 1 comp. ratio ... 40638
    Herald 1200 and 12/50 High C ... 40743
    Herald 1200 and 12/50 Low C ... 40755
    Serial numbers 25D4 ...
    High C ... 40791
    Low C ... 40790

IGNITION SYSTEM

	DM2					25D4	
	40658	40637	40638	40743	40755	40791	40790
Automatic advance commences (r.p.m.)	300	350	400	120	370	120	370
Max. advance (crankshaft degrees)	$20°$ at 2150	$17°$ at 2500	$16°$ at 2500	$10°$ at 2000	$16°$ at 2000	$10°$ at 2000	$16°$ at 2000 r.p.m.
Vacuum advance (crankshaft degrees)	$12\frac{1}{2}°$ at $11\frac{1}{2}$ in.	$10\frac{1}{2}°$ at 12 in.	$8°$ at $8\frac{1}{2}$ in.	$7\frac{1}{2}°$ at 15 in.	$12\frac{1}{2}°$ at 12 in.	$7\frac{1}{2}°$ at 15 in.	$12\frac{1}{2}°$ at 12 in. Hg.
Ignition timing	Hole in pulley and pointer on timing case

Static Timing
- 948 c.c. 7 to 1 compression ratio .... $6\frac{1}{2}°$ B.T.D.C.
- 948 c.c. 7.4 to 1 compression ratio .... $10°$ B.T.D.C.
- 948 c.c. 8 to 1 compression ratio .... $10°$ B.T.D.C.
- 948 c.c. 8.5 to 1 compression ratio .... $12°$ B.T.D.C.
- Herald 1200 early low compression engines.. $9°$ B.T.D.C.
- Herald 1200 .... $15°$ B.T.D.C.
- Herald 12/50................................ $15°$ B.T.D.C.
- Herald 13/60................................ $9°$ B.T.D.C.

NOTE. The DM2 and the 25D4 distributors are identical for all practical purposes and the few differences have no effect on the procedures used when dismantling or rebuilding either unit.

1. **GENERAL DESCRIPTION**

    In order that the engine can run correctly it is necessary for an electrical spark to ignite the fuel/air mixture in the combustion chamber at exactly the right moment in relation to engine speed and load. The ignition system is based on feeding low tension voltage from the battery to the coil where it is converted to high tension voltage. The high tension voltage is powerful enough to jump the sparking plug gap in the cylinders many times a second under high compression pressures, providing that the system is in good condition and that all adjustments are correct.

    The ignition system is divided into two circuits. The low tension circuit and the high tension circuit.

    The low tension (sometimes known as the primary) circuit consists of the battery, lead to the control box, lead to the ignition switch, lead from the ignition switch to the low tension or primary coil windings (terminal SW), and the lead from the low tension coil windings (coil terminal CB) to the contact breaker points and condenser in the distributor.

    The high tension circuit consists of the high tension or secondary coil windings, the heavy ignition lead from the centre of the coil to the centre of the distributor cap, the rotor arm, and the sparking plug leads and sparking plugs.

    The system functions in the following manner:

    Low tension voltage is changed in the coil into high tension voltage by the opening and closing of the contact breaker points in the low tension circuit. High tension voltage is then fed via the carbon brush in the centre of the distributor cap to the rotor arm of the distributor. The rotor arm revolves inside the distributor cap, and each time it comes in line with one of the four metal segments in the cap, which are connected to the sparking plug leads, the opening and closing of the contact breaker points causes the high tension voltage to build up, jump the gap from the rotor arm to the appropriate metal segment and so via the sparking plug lead to the sparking plug, where it finally jumps the spark plug gap before going to earth.

    The ignition is advanced and retarded automatically, to ensure the spark occurs at just the right instant for the particular load at the prevailing engine speed.

    The ignition advance is controlled both mechanically and by a vacuum operated system. The mechanical governor mechanism comprises two lead weights, which move out from the distributor shaft as the engine speed rises due to centrifugal force. As they move outwards they rotate the cam relative to the distributor shaft, and so advance the spark. The weights are held in position by two light springs and it is the tension of the springs which is largely responsible for correct spark advancement.

    The vacuum control consists of a diaphragm, one side of which is connected via a small bore tube to the carburetter, and the other side to the contact breaker plate. Depression in the inlet manifold and carburetter, which varies with engine speed and throttle opening, causes the diaphragm to move, so moving the contact breaker plate, and advancing or retarding the spark. A fine degree of control is achieved by a spring in the vacuum assembly.

2. **CONTACT BREAKER ADJUSTMENT**

    1. To adjust the contact breaker points to the correct gap, first pull off the two clips securing the distributor cap to the distributor body, and lift away the cap. Clean the cap inside and out with a dry cloth. It is unlikely that the four segments will be badly burned or scored, but if they are the cap will have to be renewed.

    2. Push in the carbon brush located in the top of the cap once or twice to make sure that it moves freely.

    3. Gently prise the contact breaker points open to examine the condition of their faces. If they are rough, pitted, or dirty, it will be necessary to remove them for resurfacing, or for replacement points to be fitted.

CHAPTER FOUR

4. Presuming the points are satisfactory, or that they have been cleaned and replaced, measure the gap between the points by turning the engine over until the contact breaker arm is on the peak of one of the four cam lobes (arrowed).

2.4

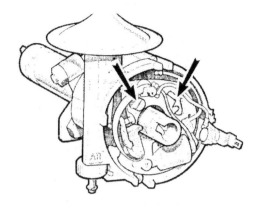

Fig. 4.1. Loosen the screw (arrowed) and by means of a screwdriver placed between the notches (arrowed) the points can be adjusted.

5. A 0.015 in. feeler gauge should now just fit between the points.
6. If the gap varies from this amount, slacken the contact plate securing screw (arrowed).

2.6

7. Adjust the contact gap by inserting a screwdriver in the notched hole (arrowed) at the end of the plate. Turning clockwise to decrease and anti-clockwise to increase the gap. Tighten the securing screw and check the gap again (small arrow).

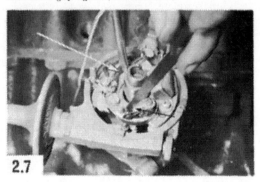

2.7

8. Replace the rotor arm and distributor cap and clip the spring blade retainers into position.

3. REMOVING & REPLACING CONTACT BREAKER POINTS

1. If the contact breaker points are burned, pitted or badly worn, they must be removed and either replaced, or their faces must be filed smooth.
2. To remove the points unscrew the terminal nut and remove it together with the steel washer under its head. Remove the flanged nylon bush and then the condenser lead and the low tension lead from the terminal pin. Lift off the contact breaker arm and then remove the large fibre washer from the terminal pin.
3. The adjustable contact breaker plate is removed by unscrewing the one holding down screw and removing it, complete with spring and flat washer.
4. To reface the points, rub their faces on a fine carborundum stone, or on fine emery paper. It is important that the faces are rubbed flat and parallel to each other so that there will be complete face to face contact when the points are closed. One of the points will be pitted and the other will have deposits on it.
5. It is necessary to completely remove the built-up deposits, but not necessary to rub the pitted point right down to the stage where all the pitting has disappeared, though obviously if this is done it will prolong the time before the operation of refacing the points has to be repeated.
6. To replace the points first position the adjustable contact breaker plate over the terminal pin (arrowed, see photograph).

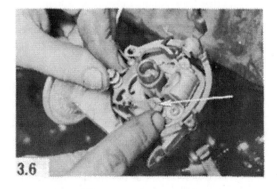

3.6

7. Secure the contact plate by screwing in the screw (arrowed) which should have a spring and a flat washer under its head.

IGNITION SYSTEM

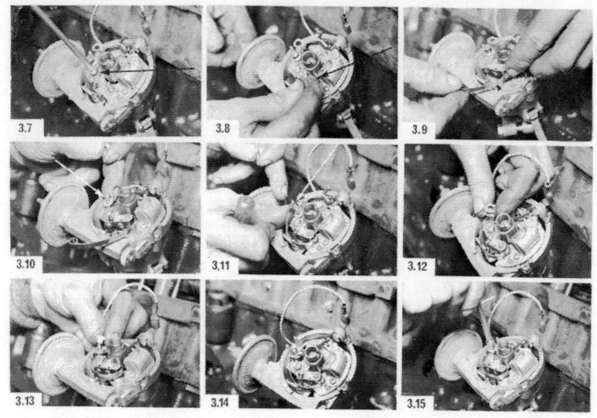

8. Then fit the fibre washer (arrowed) over the terminal pin.

9. Next fit the contact breaker arm complete with spring over the terminal pin.

10. Drop the fibre washer over the terminal bolt (arrowed).

11. Then bend back the spring of the contact breaker arm and fit it over the terminal bolt (arrowed).

12. Place the terminals of the low tension lead and the condenser over the terminal bolt.

13. Then fit the flanged nylon bush over the terminal bolt with the two leads immediately under its flange as shown.

14. Next fit a steel washer and then a 'star' washer over the nylon bush. (See photo).

15. Then fit the nut over the terminal bolt and tighten it down as shown.

16. The points are now reassembled and the gap should be set as described in the previous section.

17. Finally replace the rotor arm and then the distributor cap.

4. CONDENSER REMOVAL, TESTING & REPLACEMENT

1. The purpose of the condenser, (sometimes known as a capacitor) is to ensure that when the contact breaker points open there is no sparking across them which would waste voltage and cause wear.

2. The condenser is fitted in parallel with the contact breaker points. If it develops a short circuit, it will cause ignition failure as the points will be prevented from interrupting the low tension circuit.

3. If the engine becomes very difficult to start or begins to miss after several miles running and the breaker points show signs of excessive burning, then the condition of the condenser must be suspect. A further test can be made by separating the points by hand with the ignition switched on. If this is accompanied by a flash it is indicative that the condenser has failed.

4. Without special test equipment the only sure way to diagnose condenser trouble is to replace a suspected unit with a new one and note if there is any improvement.

5. To remove the condenser from the distributor, remove the distributor cap and the rotor arm. Unscrew the contact breaker arm terminal nut, and remove the nut, washer, and flanged nylon bush and release the condenser lead from the bush. Unscrew the condenser retaining screw from the breaker plate and remove the condenser. Replacement of the condenser is simply a reversal of the removal process. Take particular care that the condenser lead does not short circuit against any portion of the breaker plate.

5. DISTRIBUTOR LUBRICATION

1. It is important that the distributor cam is lubricated with petroleum jelly at the specified mileages, and that the breaker arm, governor weights, and cam spindle, are lubricated with engine oil once every 1,000 miles. In practice it will be found that lubrication every 3,000 miles is adequate, though once every 1,000 miles is best.

2. Great care should be taken not to use too much lubricant, as any excess that finds its way into the

CHAPTER FOUR

contact breaker points could cause burning and misfiring.

3. To gain access to the cam spindle, lift away the rotor arm. Drop no more than two drops of engine oil onto the screw head. This will run down the spindle when the engine is hot and lubricate the bearings.

4. To lubricate the automatic timing control allow a few drops of oil to pass through the hole in the contact breaker base plate through which the four sided cam emerges. Apply not more than one drop of oil to the pivot post and remove any excess.

6. **DISTRIBUTOR REMOVAL & REPLACEMENT**

1. To remove the distributor from the engine, start by pulling the terminals off each of the sparking plugs. Release the Lucar connector or small nut which holds the low tension lead to the terminal on the side of the distributor and unscrew the high tension lead retaining cap from the coil and remove the lead.

2. Unscrew the union holding the vacuum tube to the distributor vacuum housing.

3. Remove the distributor body clamp bolts which hold the distributor clamp plate to the engine and remove the distributor. NOTE if it is not wished to disturb the timing then under no circumstances should the clamp pinch bolt, which secures the distributor in its relative position in the clamp, be loosened. Providing the distributor is removed without the clamp being loosened from the distributor body, the timing will not be lost.

4. Replacement is a reversal of the above process providing that the engine has not been turned in the meantime. If the engine has been turned it will be best to retime the ignition. This will also be necessary if the clamp pinch bolt has been loosened.

7. **DISTRIBUTOR DISMANTLING**

1. With the distributor removed from the car and on the bench, remove the distributor cap and lift off the rotor arm. If very tight, lever it off gently with a screwdriver.

2. Remove the points from the distributor as described in Section 3.

3. Remove the condenser from the contact breaker plate by releasing its securing screw.

4. Unhook the vacuum unit spring from its mounting pin on the moving contact breaker plate.

5. Remove the contact breaker plate.

6. Unscrew the two screws and lockwashers which hold the contact breaker base plate in position and remove the earth lead from the relevant screw. Remember to replace this lead on reassembly.

7. Lift out the contact breaker base plate.

8. NOTE the position of the slot in the rotor arm drive in relation to the offset drive dog at the opposite end of the distributor. It is essential that this is reassembled correctly as otherwise the timing may be 180° out.

9. Unscrew the cam spindle retaining screw, which is located in the centre of the rotor arm drive, and remove the cam spindle.

10. Lift out the centrifugal weights together with their springs.

11. To remove the vacuum unit, spring off the small

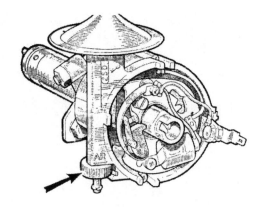

Fig. 4.2. The fine adjustment screw on the distributor.

circlip which secures the advance adjustment nut which should then be unscrewed. With the micrometer adjusting nut removed, release the spring and the micrometer adjusting nut lock spring clip. This is the clip that is responsible for the 'clicks' when the micrometer adjuster is turned, and it is small and easily lost as is the circlip, so put them in a safe place. Do not forget to replace the lock spring clip on reassembly.

12. It is only necessary to remove the distributor drive shaft or spindle if it is thought to be excessively worn. With a thin punch drive out the retaining pin from the driving tongue collar on the bottom end of the distributor drive shaft. The shaft can then be removed. The distributor is now completely dismantled.

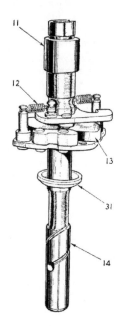

Fig.4.3. EXPLODED VIEW OF THE MAIN MOVING COMPONENTS.
11 Cam. 12 Balance weight spring. 13 Balance weight. 14 Drive shaft. 31 Collar.

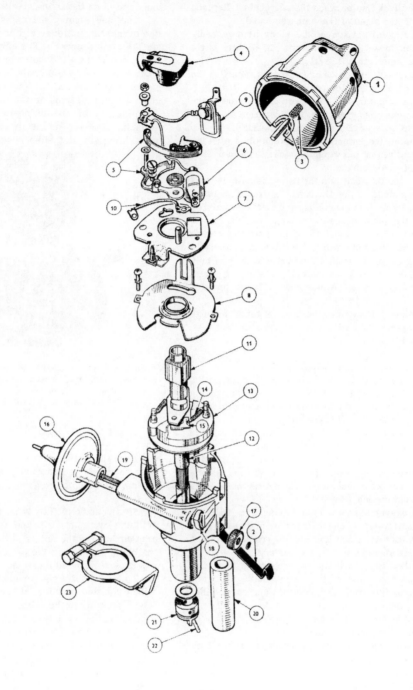

Fig. 4.4. EXPLODED VIEW OF THE DM2 DISTRIBUTOR
1 Distributor cap. 2 Cap retaining clip. 3 Cap centre brush and spring. 4 Rotor arm. 5 Contact breaker set. 6 Condenser. 7 Contact breaker base plate. 8 Contact breaker bearing plate. 9 Bush and lead assembly terminal (Lucar). 10 Earth lead. 11 Cam. 12 Shaft and action plate. 13 Automatic advance weight. 14 Automatic advance spring (set of 2). 15 Toggle. 16 Vacuum advance wrist. 17 Vacuum adjusting knurled nut. 18 Rachet spring. 19 Return spring. 20 Bushing. 21 Driving dog. 22 Securing peg for dog. 23 Clamping plate assembly.

CHAPTER FOUR

8. **DISTRIBUTOR INSPECTION & REPAIR**

1. Check the points as described in Section 3. Check the distributor cap for signs of tracking, indicated by a thin black line between the segments. Replace the cap if any signs of tracking are found.

2. If the metal portion of the rotor arm is badly burned or loose, renew the arm. If slightly burnt clean the arm with a fine file.

3. Check that the carbon brush moves freely in the centre of the distributor cover.

4. Examine the fit of the breaker plate on the bearing plate and also check the breaker arm pivot for looseness or wear and renew as necessary.

5. Examine the balance weights and pivot pins for wear, and renew the weights or cam assembly if a degree of wear is found.

6. Examine the shaft and the fit of the cam assembly on the shaft. If the clearance is excessive compare the items with new units, and renew either, or both, if they show excessive wear.

7. If the shaft is a loose fit in the distributor bush and can be seen to be worn, it will be necessary to fit a new shaft and bush. The single bush is simply pressed out. NOTE that before inserting a new bush, it should be stood in engine oil for at least 24 hours.

8. Examine the length of the balance weight springs and compare them with new springs. If they have stretched they must be renewed.

9. **DISTRIBUTOR REASSEMBLY**

1. Reassembly is a straightforward reversal of the dismantling process, but there are several points which should be noted in addition to those already given in the section on dismantling.

2. Lubricate with S.A.E. 20 engine oil the balance weights and other parts of the mechanical advance mechanism, the distributor shaft, and the portion of the shaft on which the cam bears, during assembly. Do not oil excessively but ensure these parts are adequately lubricated.

3. On reassembling the cam driving pins with the centrifugal weights, check that they are in the correct position so that when viewed from above, the rotor arm should be at six o'clock position, and the small offset on the driving dog must be on the right.

4. Check the action of the weights in the fully advanced and fully retarded positions and ensure they are not binding.

5. Tighten the micrometer adjusting nut to the middle position on the timing scale.

6. Finally, set the contact breaker gap to the correct clearance of .015 in.

10. **IGNITION TIMING**

1. If the clamp plate pinch bolt on the distributor has been loosened, or if a new or reconditioned distributor is being fitted it is necessary to set the ignition timing.

2. Turn the engine over so that No. 1 piston is coming up to T.D.C. on the compression stroke. (This can be checked by removing No. 1 sparking plug and feeling the pressure being developed in the cylinder, or by removing the rocker cover and noting when the valves in No. 4 cylinder are rocking, i.e. the inlet valve just opening and exhaust valve just closing. If this check is not made it is all too easy to set the timing 180° out, as both No. 1 and 4 cylinders come up to T.D.C. at the same time, but only one is on the firing stroke.

Continue turning the engine until the dimple on the crankshaft pulley is in line with the timing mark on the timing cover. The engine is now at T.D.C.

3. Remove the distributor cover, slacken off the distributor body clamp bolt, and with the rotor arm pointing towards the No. 1 terminal (check this position with the distributor cap and lead to No. 1 sparking plug), insert the distributor into the distributor housing. The dog on the drive shaft should match up with the slot in the distributor driving spindle.

Fig. 4.5. The rotor arm (3) should be in the position shown when No. 1. piston is at T.D.C. on its compression stroke. 1. Vernier scale. 2 Screw. 4 Clamp bolt. 5 Securing nut. 6 Washer. 7 Nut. 8 Washer. 9 Milled vernier adjuster.

Insert the two bolts holding the distributor in position.

4. Turn the knurled vernier adjuster screw clockwise to the fully retarded position indicated when the last division on the graduated scale can only just be seen.

5. Slowly and carefully turn the distributor body anti-clockwise until with the heel of the fibre rocker arm on the cam the points just begin to open. Check that with the heel on the peak of the cam the points gap does not exceed 0.016 in.

6. Tighten the distributor clamp bolt, with the distributor in this position. Check that the rotor arm is still pointing to the segment in the distributor cap which leads to No. 1 lead and plug.

7. Turn the knurled vernier adjuster screw anti-clockwise until the specified number of divisions appropriate to the model concerned are visible. Each complete division corresponds to 4 crankshaft degrees.

948 c.c. 7 to 1 C.R.	6½° BTDC	or	1¾ divisions
948 c.c. 7.4 to 1 "	10° BTDC	or	2½ divisions
948 c.c. 8 to 1 "	12° BTDC	or	3 divisions
1200 " "	15° BTDC	or	3¾ divisions
12/50 " "	15° BTDC	or	3¾ divisions
13/60 " "	9° BTDC	or	2¼ divisions

8. Difficulty is sometimes experienced in determining exactly when the contact breaker points open. This can be ascertained most accurately by connecting a 12-volt bulb in parallel with the contact breaker

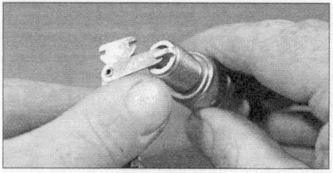

Measuring plug gap. A feeler gauge of the correct size (see ignition system specifications) should have a slight "drag" when slid between the electrodes. Adjust gap if necessary

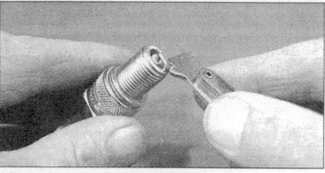

Adjusting plug gap. The plug gap is adjusted by bending the earth electrode inwards, or outwards, as necessary until the correct clearance is obtained. Note the use of the correct tool

Normal. Grey-brown deposits, lightly coated core nose. Gap increasing by around 0.001 in (0.025 mm) per 1000 miles (1600 km). Plugs ideally suited to engine, and engine in good condition

Carbon fouling. Dry, black, sooty deposits. Will cause weak spark and eventually misfire. Fault: over-rich fuel mixture. Check: carburettor mixture settings, float level and jet sizes; choke operation and cleanliness of air filter. Plugs can be re-used after cleaning

Oil fouling. Wet, oily deposits. Will cause weak spark and eventually misfire. Fault: worn bores/piston rings or valve guides; sometimes occurs (temporarily) during running-in period. Plugs can be re-used after thorough cleaning

Overheating. Electrodes have glazed appearance, core nose very white - few deposits. Fault: plug overheating. Check: plug value, ignition timing, fuel octane rating (too low) and fuel mixture (too weak). Discard plugs and cure fault immediately

Electrode damage. Electrodes burned away; core nose has burned, glazed appearance. Fault: pre-ignition. Check: as for "Overheating" but may be more severe. Discard plugs and remedy fault before piston or valve damage occurs

Split core nose (may appear initially as a crack). Damage is self-evident, but cracks will only show after cleaning. Fault: pre-ignition or wrong gap-setting technique. Check: ignition timing, cooling system, fuel octane rating (too low) and fuel mixture (too weak). Discard plugs, rectify fault immediately

IGNITION SYSTEM

points (one lead to earth and the other from the distributor low tension terminal). Switch on the ignition, and turn the advance and retard adjuster until the bulb lights up indicating that the points have just opened.

9. It must be noted that to get the very best setting the final adjustment should be made on the road. The distributor can be moved about $\frac{1}{4}$ of a division at a time until the best setting is obtained. The amount of wear in the engine, quality of petrol used, and amount of carbon in the combustion chambers all contribute to make the recommended settings no more than nominal ones. To obtain the best setting under running conditions first start the engine and allow to warm up to normal temperature, and then accelerate in top gear from 30 to 50 m.p.h. listening for heavy pinking. If this occurs, the ignition needs to be retarded slightly until just the faintest trace of pinking can be heard under these operating conditions.

10. Since the ignition advance adjustment enables the firing point to be related correctly in relation to the grade of fuel used, the fullest advantage of any change of fuel will only be attained by re-adjustment of the ignition settings.

11. SPARKING PLUGS AND LEADS

1. The correct functioning of the sparking plugs are vital for the correct running and efficiency of the engine.

2. At intervals of 6,000 miles the plugs should be removed, examined, cleaned, and if worn excessively, replaced. The condition of the sparking plug will also tell much about the overall condition of the engine.

3. If the insulator nose of the sparking plug is clean and white, with no deposits, this is indicative of a weak mixture, or too hot a plug. (A hot plug transfers heat away from the electrode slowly - a cold plug transfers it away quickly).

4. The plugs fitted as standard are the Lodge HLN or CLNY type. If the top and insulator nose is covered with hard black looking deposits, then this is indicative that the mixture is too rich. Should the plug be black and oily, then it is likely that the engine is fairly worn, as well as the mixture being too rich.

5. If the insulator nose is covered with light tan to greyish brown deposits, then the mixture is correct and it is likely that the engine is in good condition.

6. If there are any traces of long brown tapering stains on the outside of the white portion of the plug, then the plug will have to be renewed, as this shows that there is a faulty joint between the plug body and the insulator, and compression is being allowed to leak away.

7. Plugs should be cleaned by a sand blasting machine, which will free them from carbon more thoroughly than cleaning by hand. The machine will also test the condition of the plugs under compression. Any plug that fails to spark at the recommended pressure should be renewed.

8. The sparking plug gap is of considerable importance, as, if it is too large or too small, the size of the spark and its efficiency will be seriously impaired. The sparking plug gap should be set to 0.025 in. for the best results.

9. To set it, measure the gap with a feeler gauge, and then bend open, or close, the outer plug electrode until the correct gap is achieved. The centre electrode should never be bent as this may crack the insulation and cause plug failure if nothing worse.

10. When replacing the plugs, remember to use new plug washers, and replace the leads from the distributor in the correct firing order, which is 1, 3, 4, 2, No.1 cylinder being the one nearest the radiator.

11. The plug leads require no routine attention other than being kept clean and wiped over regularly. At intervals of 12,000 miles, however, pull each lead off the plug in turn and remove them from the distributor by unscrewing the knurled moulded terminal knobs. Water can seep down into these joints giving rise to a white corrosive deposit which must be carefully removed from the brass washer at the end of each cable, through which the ignition wires pass.

12. IGNITION SYSTEM FAULT - FINDING

By far the majority of breakdown and running troubles are caused by faults in the ignition system either in the low tension or high tension circuits.

13. IGNITION SYSTEM FAULT SYMPTOMS

There are two main symptoms indicating ignition faults. Either the engine will not start or fire, or the engine is difficult to start and misfires. If it is a regular misfire, i.e. the engine is only running on two or three cylinders the fault is almost sure to be in the secondary, or high tension, circuit. If the misfiring is intermittant, the fault could be in either the high or low tension circuits. If the car stops suddenly, or will not start at all, it is likely that the fault is in the low tension circuit. Loss of power and overheating, apart from faulty carburation settings, are normally due to faults in the distributor or incorrect ignition timing.

14. FAULT DIAGNOSIS - ENGINE FAILS TO START

1. If the engine fails to start and the car was running normally when it was last used, first check there is fuel in the petrol tank. If the engine turns over normally on the starter motor and the battery is evidently well charged, then the fault may be in either the high or low tension circuits. First check the H.T. circuit. NOTE: If the battery is known to be fully charged; the ignition light comes on, and the starter motor fails to turn the engine CHECK THE TIGHTNESS OF THE LEADS ON THE BATTERY TERMINALS and also the secureness of the earth lead to its CONNECTION TO THE BODY. It is quite common for the leads to have worked loose, even if they look and feel secure. If one of the battery terminal posts gets very hot when trying to work the starter motor this is a sure indication of a faulty connection to that terminal.

2. One of the commonest reasons for bad starting is wet or damp sparking plug leads and distributor. Remove the distributor cap. If condensation is visible internally dry the cap with a rag and also wipe over the leads. Replace the cap.

3. If the engine still fails to start, check that current is reaching the plugs, by disconnecting each plug lead in turn at the sparking plug end, and hold

CHAPTER FOUR

the end of the cable about 3/16 in. away from the cylinder block. Spin the engine on the starter motor by pressing the rubber button on the starter motor solenoid switch (under the bonnet).

4. Sparking between the end of the cable and the block should be fairly strong with a regular blue spark. (Hold the lead with rubber to avoid electric shocks). If current is reaching the plugs, then remove them and clean and regap them to 0.025 in. The engine should now start.

5. Spin the engine as before, when a rapid succession of blue sparks between the end of the lead and the block indicate that the coil is in order, and that either the distributor cap is cracked; the carbon brush is stuck or worn; the rotor arm is faulty, or that the contact points are burnt, pitted or dirty. If the points are in bad shape, clean and reset them as described in Section 3.

6. If there are no sparks from the end of the lead from the coil, then check the connections of the lead to the coil and distributor head, and if they are in order, check out the low tension circuit starting with the battery.

7. Switch on the ignition and turn the crankshaft so the contact breaker points have fully opened. Then with either a 20-volt voltmeter or bulb and length of wire check that current from the battery is reaching the starter solenoid switch. No reading indicates that there is a fault in the cable to the switch, or in the connections at the switch or at the battery terminals. Alternatively, the battery earth lead may not be properly earthed to the body.

8. If in order, check that current is reaching terminal 'A' (the one with the brown lead) in the control box, by connecting the voltmeter between 'A' and an earth. If there is no reading this indicates a faulty cable or loose connections between the solenoid switch and the 'A' terminal. Remedy and the car will start.

9. Check with the voltmeter between the control box terminal A1 and an earth. No reading means a fault in the control box. Fit a new control box and start the car.

10. If in order, then check that current is reaching the ignition switch by connecting the voltmeter to the ignition switch input terminal (the one connected to the brown/blue lead) and earth. No reading indicates a break in the wire or a faulty connection at the switch or A1 terminals.

11. If the correct reading (approx. 12 volts) is obtained check the output terminal on the ignition switch (the terminal connected to the white lead). No reading means that the ignition switch is broken. Replace with a new unit and start the car.

12. If current is reaching the ignition switch output terminal, then check the SW terminal on the coil (it is marked 'SW'). No reading indicates loose connections or a broken wire from the ignition switch. If this proves to be the fault, remedy and start the car.

13. Check the CB terminal on the coil (it is marked 'CB') and if no reading is recorded on the voltmeter then the coil is broken and must be replaced. The car should start when a new coil has been fitted.

14. If a reading is obtained at the CB terminal then check the low tension terminal on the side of the distributor. If no reading then check the wire for loose connections etc. If a reading is obtained then the final check on the low tension is across the breaker points. No reading means a broken condenser which when replaced will enable the car to finally start.

15. FAULT DIAGNOSIS - ENGINE MISFIRES

1. If the engine misfires regularly, run it at a fast idling speed, and short out each of the plugs in turn by placing a short screwdriver across from the plug terminal to the cylinder. Ensure that the screwdriver has a WOODEN or PLASTIC INSULATED HANDLE.

2. No difference in engine running will be noticed when the plug in the defective cylinder is short circuited. Short circuiting the working plugs will accentuate the misfire.

3. Remove the plug lead from the end of the defective plug and hold it about 3/16 in. away from the block. Restart the engine. If the sparking is fairly strong and regular the fault must lie in the sparking plug.

4. The plug may be loose, the insulation may be cracked, or the points may have burnt away giving too wide a gap for the spark to jump. Worse still, one of the points may have broken off. Either renew the plug, or clean it, reset the gap, and then test it.

5. If there is no spark at the end of the plug lead, or if it is weak and intermittent, check the ignition lead from the distributor to the plug. If the insulation is cracked or perished, renew the lead. Check the connections at the distributor cap.

6. If there is still no spark, examine the distributor cap carefully for tracking. This can be recognised by a very thin black line running between two or more electrodes, or between an electrode and some other part of the distributor. These lines are paths which now conduct electricity across the cap thus letting it run to earth. The only answer is a new distributor cap.

7. Apart from the ignition timing being incorrect, other causes of misfiring have already been dealt with under the section dealing with the failure of the engine to start. (Section 14).

8. If the ignition timing is too far retarded, it should be noted that the engine will tend to overheat and there will be a quite noticeable drop in power. If the engine is overheating and the power is down, and the ignition timing is correct, then the carburetters should be checked, as it is likely that this is where the fault lies. See Chapter 3 for details on this.

CHAPTER FIVE

CLUTCH AND ACTUATING MECHANISM

CONTENTS

General Description...	1
Routine Maintenance...	2
Clutch System - Bleeding...	3
Clutch Pedal - Removal & Replacement...	4
Clutch - Removal...	5
Clutch - Replacement...	6
Clutch - Dismantling...	7
Clutch - Inspection...	8
Clutch - Reassembly...	9
Clutch Slave Cylinder - Removal, Dismantling, Examination & Reassembly...	10
Clutch Master Cylinder - Removal, Dismantling, Examination & Reassembly...	11
Clutch Release Bearing - Removal & Replacement...	12
Clutch Release Bearing - Adjustment...	13
Clutch Faults...	14
Clutch Squeal - Diagnosis & Cure...	15
Clutch Slip - Diagnosis & Cure...	16
Clutch Spin - Diagnosis & Cure...	17

SPECIFICATIONS

Up to Herald 1200 Engine Nos. GA.204020E and GB.24121 and Herald 12/50 GD.44446E

Make	Borg & Beck
Type	Coil springs - single dry plate
Diameter	$6\frac{1}{4}$ in. (15.87 cm.)
Pressure springs	6
Damper springs	4
Clutch facing	Mintex M19
Clutch release bearing	Single row ball bearing
Clutch fluid	Lockheed or Girling hydraulic fluid

From Herald 1200 Engine Nos. GA.204020E and GB.24121 and Herald 12/50 GD.44446E and all Herald 13/60

Make	Borg & Beck
Type	Diaphragm spring
Diameter	$6\frac{1}{2}$ in. (16.51 cm.)
Number of pressure springs	One (diaphragm spring)
Number of damper springs	4
Damper spring colour	White/light green
Clutch facing material	Mintex M19
Clutch release bearing	Single row ball bearing
Clutch fluid	Lockheed or Girling hydraulic fluid

TORQUE WRENCH SETTINGS

Clutch housing to gearbox	24 to 26 lbs/ft. (3.318 to 3.595 kgs.m.)
Slave cylinder to mounting bracket	10 to 12 lbs/ft. (1.383 to 1.659 kgs.m.)

1. **GENERAL DESCRIPTION**

Early models are fitted with a Borg & Beck single dry plate clutch of $6\frac{1}{4}$ in. diameter while later models are fitted with a $6\frac{1}{2}$ in. diaphragm spring unit.

The $6\frac{1}{4}$ in. clutch comprises a steel cover which is bolted and dowelled to the rear face of the flywheel and contains the pressure plate, pressure

CLUTCH AND ACTUATING MECHANISM

plate springs, release levers, and clutch disc or driven plate. The layout of the clutch is shown in Fig. 5.1.

The pressure plate, pressure springs, and release levers are all attached to the clutch assembly cover. The clutch disc is free to slide along the splined first motion shaft and is held in position between the flywheel and the pressure plate by the pressure of the pressure plate springs.

Friction lining material is riveted to the clutch disc and it has a spring cushioned hub to absorb transmission shocks and to help ensure a smooth take-off.

The $6\frac{1}{2}$ in. diaphragm spring clutch is very similar to the $6\frac{1}{4}$ in. unit but there are two main alterations. In place of the coil pressure springs there is just one diaphragm spring and this dispenses with the need for release levers. (See Fig. 5.2.).

The clutch is actuated hydraulically. The pendant clutch pedal, is connected to the clutch master cylinder and hydraulic fluid reservoir by a short push rod. The master cylinder and hydraulic reservoir are mounted on the engine side of the bulkhead in front of the driver.

Depressing the clutch pedal moves the piston in the master cylinder forwards, so forcing hydraulic fluid through the clutch hydraulic pipe to the slave cylinder.

The piston in the slave cylinder moves forward on the entry of the fluid and actuates the clutch release arm by means of a short pushrod.

On $6\frac{1}{4}$ in. clutches the release bearing is pushed forwards to bear against the release bearing thrust plate and three clutch release levers. These levers are pivoted so as to move the pressure plate backwards against the pressure of the pressure plate springs, in this way disengaging the pressure plate from the clutch disc.

When the clutch pedal is released, the pressure plate springs force the pressure plate into contact with the high friction linings on the clutch disc, at the same time forcing the clutch disc against the flywheel and so taking up the drive.

On models fitted with the diaphragm spring clutch the release arm pushes the release bearing forwards to bear against the release plate, so moving the centre of the diaphragm spring inwards. The spring is sandwiched between two annular rings which act as fulcrum points. As the centre of the spring is pushed in the outside of the spring is pushed out, so moving the pressure plate backwards and disengaging the pressure plate from the clutch disc.

When the clutch pedal is released the diaphragm spring forces the pressure plate into contact with the high friction linings on the clutch disc and at the same time pushes the clutch disc a fraction of an inch forwards on its splines so engaging the clutch disc with the flywheel. The clutch disc is now firmly sandwiched between the pressure plate and the flywheel so the drive is taken up.

As the friction linings on the clutch disc wear the pressure plate automatically moves closer to the disc to compensate. There is therefore no need to periodically adjust either type of clutch.

2. ROUTINE MAINTENANCE
1. Routine maintenance consists of checking the level of the hydraulic fluid in the master cylinder every 1,000 miles and topping up with Lockheed or Girling hydraulic fluid if the level falls.
2. If it is noted that the level of the liquid has fallen then an immediate check should be made to determine the source of the leak.
3. Before checking the level of the fluid in the master cylinder reservoir, carefully clean the cap and body of the reservoir unit with clean rag so as to ensure that no dirt enters the system when the cap is removed. On no account should paraffin or any other cleaning solvent be used in case the hydraulic fluid becomes contaminated.
4. Check that the level of the hydraulic fluid is up to within $\frac{1}{4}$ in. of the filler neck and that the vent hole in the cap is clear. Do not overfill.

3. CLUTCH SYSTEM - BLEEDING
1. Gather together a clean jam jar, a 9 in. length of rubber tubing which fits tightly over the bleed nipple in the slave cylinder, a tin of hydraulic brake fluid, and a friend to help.
2. Check that the master cylinder is full and if not fill it, and cover the bottom two inches of the jar with hydraulic fluid.
3. Remove the rubber dust cap from the bleed nipple on the slave cylinder (arrowed), and with a suitable spanner open the bleed nipple one turn.

3.3

4. Place one end of the tube securely over the nipple and insert the other end in the jam jar so that the tube orifice is below the level of the fluid.
5. The assistant should now pump the clutch pedal up and down slowly until air bubbles cease to emerge from the end of the tubing. He should also check the reservoir frequently to ensure that the hydraulic fluid does not disappear so letting air into the system.
6. When no more air bubbles appear, tighten the bleed nipple on the downstroke.
7. Replace the rubber dust cap over the bleed nipple. Allow the hydraulic fluid in the jar to stand for at least 24 hours before using it, to allow all the minute air bubbles to escape.

4. CLUTCH PEDAL - REMOVAL & REPLACEMENT
1. The clutch pedal is removed and replaced in exactly the same way as the brake pedal.
2. A full description of how to remove and replace the brake pedal can be found in Chapter 9/13.

CHAPTER FIVE

5. **CLUTCH REMOVAL**

1. Remove the gearbox as described in Chapter 6, Section 3.
2. Remove the clutch assembly by unscrewing the six bolts holding the cover to the rear face of the flywheel (photo). Unscrew the bolts diagonally half a turn at a time to prevent distortion to the cover flange.

3. With all the bolts and spring washers removed lift the clutch assembly off the locating dowels. The driven plate or clutch disc will fall out at this stage as it is not attached to either the clutch cover assembly or the flywheel (photo).

6. **CLUTCH REPLACEMENT**

1. It is important that no oil or grease gets on the clutch disc friction linings, or the pressure plate and flywheel faces. It is advisable to replace the clutch with clean hands and to wipe down the pressure plate and flywheel faces with a clean dry rag before assembly begins.
2. Place the clutch disc against the flywheel with the longer end of the hub facing outwards (photo), away from the flywheel. On no account should the clutch disc be replaced with the longer end of the centre hub facing in to the flywheel as on reassembly it will be found quite impossible to operate the clutch with the friction disc in this position.
3. Replace the clutch cover assembly loosely on the dowels, (one dowel is arrowed in the photo). Replace the six bolts and spring washers and tighten them finger tight so that the clutch disc is gripped but can still be moved.

4. The clutch disc must now be centralised so that when the engine and gearbox are mated, the gearbox input shaft splines will pass through the splines in the centre of the driven plate hub.
5. Centralisation can be carried out quite easily by inserting a round bar or long screwdriver through the hole in the centre of the clutch, so that the end of the bar rests in the small hole in the end of the crankshaft containing the input shaft bearing bush. Ideally an old Triumph input shaft should be used.
6. Using the input shaft bearing bush as a fulcrum, moving the bar sideways or up and down will move the clutch disc in whichever direction is necessary to achieve centralisation.
7. Centralisation is easily judged by removing the bar and viewing the driven plate hub in relation to the hole in the release bearing. When the hub appears exactly in the centre of the release bearing hole all is correct. Alternatively the input shaft, (arrowed) will fit the bush and centre of the clutch hub exactly, obviating the need for visual alignment (photo).

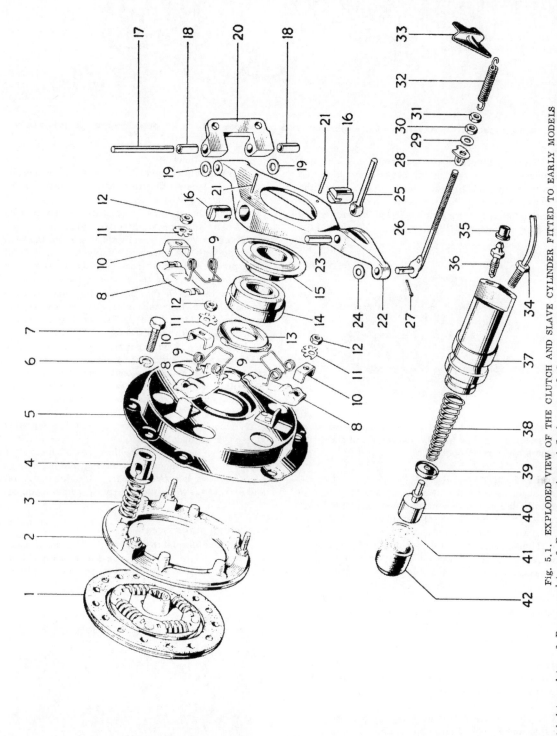

Fig. 5.1. EXPLODED VIEW OF THE CLUTCH AND SLAVE CYLINDER FITTED TO EARLY MODELS

1 Clutch driven plate. 2 Pressure plate. 3 Pressure spring. 4 Spring cup. 5 Cover pressing. 6 Spring washer. 7 Setscrew. 8 Release levers. 9 Anti-rattle spring. 10 Bridge pieces. 11 Locking plates. 12 Adjusting nuts. 13 Release lever plate. 14 Release bearing. 15 Release bearing carrier. 16 Thrust plugs. 17 Hinge pin. 18 Bushes. 19 Distance washer. 20 Hinge plate. 21 Pin. 22 Crossmember. 23 Pin. 24 Plain washer. 25 Pushrod. 26 Adjusting rod. 27 Split pin. 28 Abutment bracket. 29 Plain washer. 30 Locknut. 31 Adjusting nut. 32 Pull-off spring. 33 Spring anchor bracket. 34 Pressure pipe. 35 Dust cap. 36 Bleed nipple. 37 Slave cylinder. 38 Return spring. 39 Sealing cup. 40 Piston. 41 Circlip. 42 Dust excluder.

CHAPTER FIVE

Fig. 5.3. If the proper tool is not available use a hydraulic press and wooden blocks to compress the pressure springs as shown.

8. Tighten the clutch bolts firmly in a diagonal sequence to ensure that the cover plate is pulled down evenly, (photo) and without distortion of the flange. Note how the flywheel is prevented from turning by a spanner located between the teeth of the starter ring and a bellhousing stud.

9. Mate the engine and gearbox, bleed the slave cylinder if the pipe was disconnected and check the clutch for correct operation.

7. CLUTCH DISMANTLING

1. It is not very often that it is necessary to dismantle the clutch cover assembly, and in the normal course of events clutch replacement is the term used for simply fitting a new clutch disc. Under no circumstances must the diaphragm clutch unit be dismantled. If a fault develops in the unit an exchange replacement assembly must be fitted.
2. If a new clutch disc is being fitted it is a false economy not to renew the release bearing at the same time. This will preclude having to replace it at a later date when wear on the clutch linings is still very small.
3. It should be noted here that it is preferable to purchase an exchange clutch cover assembly unit, which will have been properly balanced rather than to dismantle and repair the existing cover.
4. Before beginning work ensure that either the Churchill clutch assembly gauging tool 99A or a press and a block of wood is available for compressing the clutch springs so that the three adjusting nuts can be freed.
5. Presuming that it is possible to borrow from your local Triumph agent, clutch assembly tool 99A,

proceed as follows:-
6. Mark the clutch cover, release levers, and pressure plate lugs so that they can be refitted in the same relative positions.
7. Unhook the springs from the release bearing thrust plate and remove the plate and spring.
8. Place the three correctly sized spacing washers provided with the clutch assembly tool on the tool base plate in the positions indicated by the chart (found inside the lid of the assembly tool container).
9. Place the clutch face down on the three spacing washers so that the washers are as close as possible to the release levers, with the six holes in the cover flange in line with the six holes in the base plate.
10. Insert the six bolts provided with the assembly tool through the six holes in the cover flange, and tighten the cover down diagonally onto the base plate.
11. With a suitable punch, tap back the three tab washers and then remove the three adjusting nuts and bearing plates from the pressure plate bolts on early models, and just unscrew the three adjusting nuts on later models.
12. Unscrew the six bolts holding the clutch cover to the base plate, diagonally, and a turn at a time, so as to release the cover evenly. Lift the cover off and extract the six pressure spring retaining cups.

8. CLUTCH INSPECTION

1. Examine the clutch disc friction linings for wear and loose rivets and the disc for rim distortion, cracks, broken hub springs, and worn splines. Shown is a well worn friction plate.

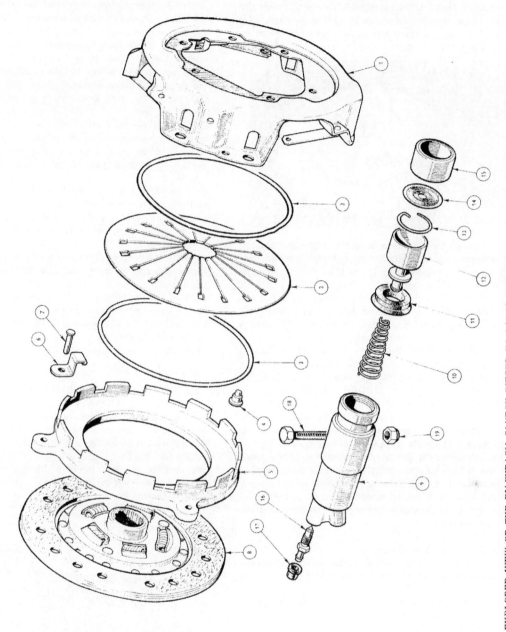

Fig. 5.2 EXPLODED VIEW OF THE DIAPHRAGM CLUTCH FITTED TO LATER MODELS AND THE HYDRAULIC SLAVE CYLINDER

1 **Cover and straps.** 2 Diaphragm spring. 3 Diaphragm spring fulcrum ring. 4 Rivet. 5 Pressure plate. 6 Retaining clip. 7 Rivet (holds cover strap to pressure plate). 8 **Clutch disc or driven plate.** 9 Slave cylinder body. 10 Spring. 11 Piston seal. 12 Piston. 13 Circlip. 14 Dust cover. 15 Dust cover retainer. 16 Bleed screw. 17 Bleed screw cover. 18 Bolt. 19 Nyloc nut.

2. It is always best to renew the clutch driven plate as an assembly to preclude further trouble, but, if it is wished to merely renew the linings, the rivets should be drilled out and not knocked out with a punch. The manufacturers do not advise that only the linings are renewed and personal experience dictates that it is far more satisfactory to renew the driven plate complete than to try and economise by only fitting new friction linings. Shown in the photo is a new friction plate.

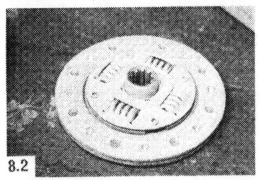
8.2

3. Check the machined faces of the flywheel and the pressure plate. If either are badly grooved they should be machined until smooth. If the pressure plate is cracked or split it must be renewed, also if the portion on the other side of the plate in contact with the three release lever tips are grooved.
4. Check the release bearing thrust plate for cracks and renew it if any are found.
5. Examine the tips of the release levers which bear against the thrust plate, and renew the levers if more than a small flat has been worn on them.
6. Renew any clutch pressure springs that are broken or shorter than standard.
7. Examine the depressions in the release levers which fit over the knife edge fulcrums and renew the levers if the metal appears badly worn.
8. Examine the clutch release bearing in the gearbox bellhousing and if it turns or if it is roughly cracked or pitted, it must be removed and replaced.
9. Also check the clutch withdrawal lever for slackness. If this is evident, withdraw the lever and renew the bush.

9. CLUTCH REASSEMBLY

1. During clutch reassembly ensure that the marked components are placed in their correct relative positions.
2. Place the three spacing washers on the clutch assembly tool base in the same position as for dismantling the clutch.
3. Place the clutch pressure plate face down on the three spacing washers.
4. Position the three release levers on the knife edge fulcrums (or release lever floating pins in the later clutches) and ensure that the anti-rattle springs are in place over the inner end of the levers.
5. Position the pressure springs on the pressure plate bosses.
6. Fit the flanged cups to the clutch cover and fit the cover over the pressure plate in the same relative position as it was originally.
7. Insert the six assembly tool bolts through the six holes in the clutch cover flange and tighten the cover down, diagonally, a turn at a time.
8. Replace the three bearing plates, tag washers, and adjusting nuts over the pressure plate studs in the early units, and just screw the adjusting nuts into the eyebolts in the later models.
9. To correctly adjust the clutch release levers use the clutch assembly tool as detailed below:-
a) Screw the actuater into the base plate and settle the clutch mechanism by pumping the actuater handle up and down a dozen times. Unscrew the actuater.
b) Screw the tool piller into the base plate and slide the correctly sized distance piece (as indicated in the chart in the tool's box) recessed side downwards, over the pillar.
c) Slip the height finger over the centre pillar and turn the release lever adjusting nuts, until the height fingers, when rotated and held firmly down, just contact the highest part of the clutch release lever tips.
d) Remove the pillar, replace the actuater, and settle the clutch mechanism as in (a).
e) Refit the centre pillar and height finger and re-check the clutch release lever clearance, and adjust if not correct.
10. With the centre pillar removed, lock the adjusting nuts found on early clutches by bending up the tab washers.
11. Replace the release bearing thrust plate and fit the retaining springs over the thrust plate hooks.
12. Unscrew the six bolts holding the clutch cover to the base plate, diagonally, a turn at a time and assembly is now complete.

10. CLUTCH SLAVE CYLINDER - REMOVAL, DISMANTLING, EXAMINATION & REASSEMBLY

1. The clutch slave cylinder is positioned on the left-hand side of the bellhousing.
2. Before removing the cylinder take off the clutch reservoir cap and place a piece of thin polythene over the top of the reservoir. Screw down the cap tightly. Then pull off the slave cylinder release spring (photo).

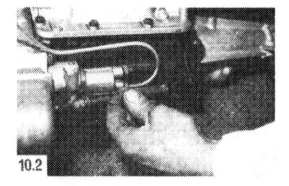
10.2

3. Undo the vertical clamp bolt and nut (photo). disconnect the hydraulic pipe at the cylinder (the polythene under the cap will stop the fluid leaking) and pull out the slave cylinder.

CLUTCH AND ACTUATING MECHANISM

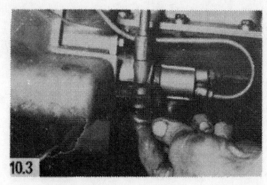

1. Referring to Fig. 5.4. pull off the rubber cover (32), and take out the circlip (31). The piston (30), seal (29), and spring (28) can then be shaken from the slave cylinder bore. Clean all the components thoroughly with hydraulic fluid or alcohol and dry them off.

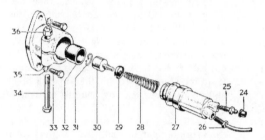

Fig. 5.4. Exploded view of the slave cylinder. The numbers refer to parts mentioned in the text.

5. Carefully examine the rubber components for signs of swelling, distortion, splitting or other wear, and check the piston and cylinder wall for wear and score marks. Replace any parts that are found faulty.

6. Reassembly is a straightforward reversal of the dismantling procedure, but NOTE the following points:-

a) As the component parts are refitted to the slave cylinder barrel, smear them with hydraulic fluid.

b) When reassembling the operating piston, locate the piston seal at the end of the piston so that the sealing lip is towards the closed end of the slave cylinder bore.

c) On completion of reassembly, top up the reservoir tank with the correct grade of hydraulic fluid and bleed the system, if necessary.

11. **CLUTCH MASTER CYLINDER - REMOVAL, DISMANTLING, EXAMINATION & REASSEMBLY**

1. Referring to Fig. 5.5. pull back the rubber dust excluder (11) and remove the split pin (12) from the clevis (14) which holds the pushrod to the top of the brake pedal (8). Pull out the clevis pin (14).

2. Place a rag under the master cylinder to catch any hydraulic fluid which may be spilt. Unscrew the union nut from the end of the hydraulic pipe where it enters the clutch master cylinder and gently pull the pipe clear.

3. Unscrew the two bolts and spring washers holding the clutch cylinder mounting flange to the mounting bracket.

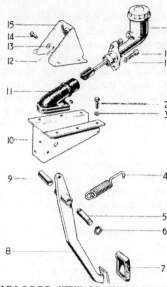

Fig. 5.5. EXPLODED VIEW OF THE CLUTCH PEDAL & BRACKET ASSEMBLY

1 Master cylinder. 2 Bolt. 3 Spring washer. 4 Return spring. 5 Pivot pin. 6 Circlip. 7 Pedal rubber. 8 Pedal. 9 Pedal pivot bush. 10 Pedal bracket. 11 Rubber dust excluder. 12 Split pin. 13 Plain washer. 14 Clevis pin. 15 Master cylinder bracket. 16 Bolt. 17 Spring washer.

4. Remove the master cylinder and reservoir, unscrew the filler cap, and drain the hydraulic fluid into a clean container.

5. Referring to Fig.5.6. pull off the rubber boot (10) to expose the circlip (11) which must be removed so the pushrod complete with metal retaining washer (12) can be pulled out of the master cylinder (14).

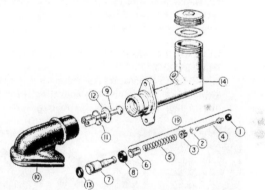

Fig. 5.6. EXPLODED VIEW OF THE CLUTCH MASTER CYLINDER

1 Valve seal. 2 Dished seal washer. 3 Spacer. 4 Valve shank. 5 Spring. 6 Spring retainer. 7 Piston. 8 Piston seal. 9 Pushrod. 10 Boot. 11 Circlip. 12 Washer. 13 Secondary piston seal.

6. With a small electrical screwdriver lift the tag on the spring retainer (6) which engages against the shoulder on the front of the piston shank (7) and separate the piston from the retainer.

7. To dismantle the valve assembly manoeuvre the flange on the valve shank stem (4) through the eccentrically positioned hole in the end face of the spring retainer (6). The spring (5), distance piece (3) and valve spring seal washer (2) can now be pulled off the valve shank stem (4)

8. Carefully ease the rubber seals (1, 8, & 13) from the valve stem (4) and the piston (7) respectively.

CHAPTER FIVE

9. Clean and carefully examine all the parts, especially the piston cup and rubber washers, for signs of distortion, swelling, splitting, or other wear and check the piston and cylinder for wear and scoring. Replace any parts that are faulty.

10. During the inspection of the piston seal it has been found advisable to maintain the shape of this seal as regular as possible and for this reason do not turn it inside out as slight distortion may be caused.

11. Rebuild the piston and valve assembly in the following sequence.

a) Fit the piston seal (8) to the piston (7) so the larger circumference of the rubber lip will enter the cylinder bore first. Then fit the seal (13) in the same manner.

b) Fit the valve seal (1) to the valve (4) in the same way.

c) Place the valve spring seal washer (2) so its convex face abuts the valve stem flange, and then fit the seat spacer (3) and spring (5).

d) Fit the spring retainer (6) to the spring (5) which must then be compressed so the valve stem (4) can be reinserted in the retainer (6).

e) Replace the front of the piston (7) in the retainer (6) and then press down the retaining leg so it locates under the shoulder at the front of the piston shank.

f) Generously lubricate the assembly with hydraulic fluid and carefully replace it in the master cylinder taking great care not to damage the rubber seals as they are inserted into the cylinder bore.

g) Fit the pushrod (9) and washer (12) in place and secure with the circlip (11). Replace the rubber boot (10).

12. Replacement of the unit in the car is a straightforward reversal of the removal sequence. Finally, bleed the system as described earlier in Section 3.

12. CLUTCH RELEASE BEARING - REMOVAL & REPLACEMENT

1. All figures in brackets refer to Fig 5.1. With a thin metal drift drive the operating lever fulcrum pin (17) out of the bellhousing (photo).

2. Remove the operating lever (22) complete with the release bearing and sleeve (14,15), (photo).

3. Place the operating lever on top of a vice and with the aid of a sawn off nail drive out the pins (21) (photo).

4. The retaining plugs (16) can then be partially levered out (photo) so as to release the bearing sleeve (15) and bearing (14).

5. Place the old bearing in a vice and carefully lever off the bearing sleeve (photo).

6. NOTE that the raised edge of the bearing is away from the sleeve and with the aid of a block of wood and a vice press the new bearing onto the old sleeve (photo).

7. Refit the sleeve to the operating arm, tap back the retaining plugs, and refit the pins (photo). Replace the arm in the bellhousing, and finally drift in the pin and lightly stake it in place.

13. CLUTCH RELEASE BEARING - ADJUSTMENT

1. On early 948 c.c. models provision is made for adjustment of the release bearing. As the clutch linings wear the release lever plate moves backwards so reducing the clearance between the plate and the release bearing.

2. Referring to Fig 5.7, unhook the return spring (5), loosen the locknut (2) and undo the adjusting nut (3) so the rod (1) is free.

3. The rod must then be pushed into the clutch

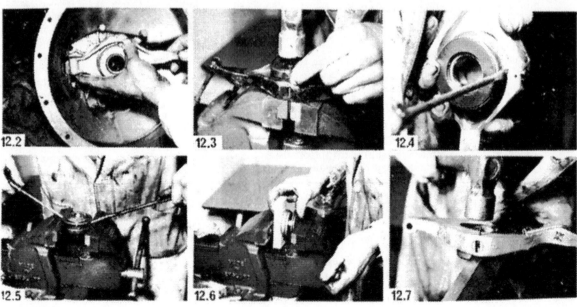

CLUTCH AND ACTUATING MECHANISM

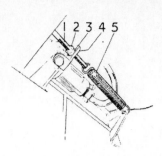

Fig. 5.7. CLUTCH ADJUSTMENT
1 Adjusting rod. 2 Locknut. 3 Adjusting nut. 4 Abutment bracket. 5 Return spring.

housing and the nuts (2, 3) turned until a 0.080 in. (2 mm.) feeler blade just slips in between the nut (3) and the abutment (4). Tighten the locknut, taking care not to turn the adjustment nut, and replace the spring.

14. CLUTCH FAULTS

There are four main faults to which the clutch and release mechanism are prone. They may occur by themselves or in conjunction with any of the other faults. They are clutch squeal, slip, spin, and judder.

15. CLUTCH SQUEAL - DIAGNOSIS & CURE

1. If on taking up the drive or when changing gear, the clutch squeals, this is a sure indication of a badly worn clutch release bearing.
2. As well as regular wear due to normal use, wear of the clutch release bearing is much accentuated if the clutch is ridden, or held down for long periods in gear, with the engine running. To minimise wear of this component the car should always be taken out of gear at traffic lights and for similar hold-ups.
3. The clutch release bearing is not an expensive item, but difficult to get at.

16. CLUTCH SLIP - DIAGNOSIS & CURE

1. Clutch slip is a self-evident condition which occurs when the clutch friction plate is badly worn, the release arm free travel is insufficient, oil or grease have got onto the flywheel or pressure plate faces, or the pressure plate itself is faulty.
2. The reason for clutch slip is that, due to one of the faults listed above, there is either insufficient pressure from the pressure plate, or insufficient friction from the friction plate to ensure solid drive.
3. If small amounts of oil get onto the clutch, they will be burnt off under the heat of clutch engagement, and in the process, gradually darkening the linings. Excessive oil on the clutch will burn off leaving a carbon deposit which can cause quite bad slip, or fierceness, spin and judder.
4. If clutch slip is suspected, and confirmation of this condition is required, there are several tests which can be made.

5. With the engine in second or third gear and pulling lightly up a moderate incline, sudden depression of the accelerator pedal may cause the engine to increase its speed without any increase in road speed. Easing off on the accelerator will then give a definite drop in engine speed without the car slowing.
6. In extreme cases of clutch slip the engine will race under normal acceleration conditions.
7. If slip is due to oil or grease on the linings a temporary cure can sometimes be effected by squirting carbon tetrachloride into the clutch. The permanent cure is, of course, to renew the clutch driven plate and trace and rectify the oil leak.

17. CLUTCH SPIN - DIAGNOSIS & CURE

1. Clutch spin is a condition which occurs when there is a leak in the clutch hydraulic actuating mechanism, the release arm free travel is excessive, there is an obstruction in the clutch either on the primary gear splines, or in the operating lever itself, or the oil may have partially burnt off the clutch linings and have left a resinous deposit which is causing the clutch disc to stick to the pressure plate or flywheel.
2. The reason for clutch spin is that due to any, or a combination of, the faults just listed, the clutch pressure plate is not completely freeing from the centre plate even with the clutch pedal fully depressed.
3. If clutch spin is suspected, the condition can be confirmed by extreme difficulty in engaging first gear from rest, difficulty in changing gear, and very sudden take-up of the clutch drive at the fully depressed end of the clutch pedal travel as the clutch is released.
4. Check the clutch master and slave cylinders and the connecting hydraulic pipe for leaks. Fluid in one of the rubber boots fitted over the end of either the master or slave cylinders is a sure sign of a leaking piston seal.
5. If these points are checked and found to be in order then the fault lies internally in the clutch, and it will be necessary to remove the clutch for examination.

18. CLUTCH JUDDER - DIAGNOSIS & CURE

1. Clutch judder is a self-evident condition which occurs when the gearbox or engine mountings are loose or too flexible, when there is oil on the faces of the clutch friction plate, or when the clutch pressure plate has been incorrectly adjusted.
2. The reason for clutch judder is that due to one of the faults just listed, the clutch pressure plate is not freeing smoothly from the friction disc, and is snatching.
3. Clutch judder normally occurs when the clutch pedal is released in first or reverse gears, and the whole car shudders as it moves backwards or forwards.

CHAPTER SIX
GEARBOX

CONTENTS

General Description	1
Routine Maintenance	2
Gearbox - Removal & Replacement	3
Gearbox - Dismantling	4
Gearbox - Examination & Renovation	5
Input Shaft - Dismantling & Reassembly	6
Mainshaft - Dismantling & Reassembly	7
Gearbox - Reassembly	8
Gear Selectors - Removal & Replacement	9
Remote Control Assembly - Overhaul	10

SPECIFICATIONS

Gearbox

No. of gears	4 forward, 1 reverse.
Synchromesh	2nd, 3rd and 4th.
Laygear endfloat	.0015 to .0125 in. (.04 to .31 mm.).
Mainshaft 2nd & 3rd gear endfloat	.002 to .006 in. (.05 to .1524 mm.).
Oil capacity	1.5 pints (1.8 U.S. pints, .85 litres).
Oil type	S.A.E. 90 E.P. gear oil.

Gearbox Ratios

	948 c.c. Heralds	Herald 1200, 12/50 & 13/60
First	4.271 to 1	3.75 to 1
Second	2.46 to 1	2.16 to 1
Third	1.454 to 1	1.40 to 1
Fourth	1.00 to 1	1.00 to 1
Reverse	4.271 to 1	3.75 to 1

Overall Ratios

	948 c.c. Saloons	948 c.c. Coupes	Herald 1200, 12/50 & 13/60
First	20.82 to 1	19.45 to 1	15.4 to 1
Second	11.99 to 1	11.2 to 1	8.88 to 1
Third	7.09 to 1	6.62 to 1	5.74 to 1
Fourth	4.875 to 1	4.55 to 1	4.11 to 1

Synchromesh Hub Springs

Axial release load	19 to 21 lbs. (8.618 to 9.525 kg).

TORQUE WRENCH SETTINGS

Layshaft location bolt	24 to 26 lb/ft. (3.318 to 3.595 kg.m).
Gearbox extension set screws	14 to 16 lb/ft. (1.936 to 2.212 kg.m).
Extension to top cover studs	12 to 14 lb/ft. (1.659 to 1.936 kg.m).
Flange to mainshaft	70 to 80 lb/ft. (9.678 to 11.060 kg.m).
Operating shaft to gear lever	6 to 8 lb/ft. (0.830 to 1.106 kg.m).
Reverse idler shaft	14 to 16 lb/ft. (1.936 to 2.212 kg.m).
Speedometer sleeve attachment	14 to 16 lb/ft. (1.936 to 2.212 kg.m).
Top cover attachment	6 to 8 lb/ft. (0.830 to 1.106 kg.m).

GEARBOX

1. GENERAL DESCRIPTION

The gearbox fitted to all models contains four forward and one reverse gear. Synchromesh is fitted between second and third, and third and fourth gears. The aluminium alloy bellhousing is a separate casting to the cast iron gearbox casing. Attached to the rear of the gearbox is an extension which houses the tail end of the mainshaft. Gears are selected by means of a remote control assembly, the casing for which is bolted to the top of the gearbox.

2. ROUTINE MAINTENANCE

Once every 6,000 miles on pre-1965 models check the level of oil in the gearbox by removing the oil filler/level plug on the right-hand side of the gearbox. Top up, if necessary, with SAE 90 EP gear oil until the oil starts to run out of the filler hole.

On post 1965 models check the level of oil in the gearbox once every 12,000 miles. On early models when the oil is hot, drain and refill the gearbox at this mileage. On later models no specific recommendations are made by the manufacturer. In the Author's experience it is best to change the oil at least every 24,000 miles, as in this way the majority of abrasive metal particles which are bound to accumulate as the mileage rises, are carried away with the old oil.

3. GEARBOX - REMOVAL & REPLACEMENT

1. The gearbox can be removed in unit with the engine as described in Chapter 1/7. Alternatively, the better method is to separate the gearbox bellhousing from the engine end plate and to lift the gearbox out from inside the car, leaving the engine in position.
2. Drain the gearbox oil, take out the carpets, and disconnect the battery.
3. Where fitted, undo the bolts which hold the fascia support which straddles the end of the gearbox cover in place and disconnect the tachometer drive. Undo the fasteners and the three screws on the engine side of the bulkhead which holds the gearbox cover in place (photo).

4. Loosen the gearlever knob locknut and then screw the knob off the gearlever (photo). Remove the gearbox cover from inside the car.

5. Undo the knurled nut which holds the speedometer drive cable to the gearbox extension (photo).
6. Some models make use of a special stay to restrict the fore and aft movement of the power unit (photo).
7. Where this stay is fitted, grip it firmly with a mole wrench to prevent it turning and undo the nut holding it to the gearbox (photo).
8. Scratch a mating mark across the propeller shaft front flange and the mainshaft drive flange and then undo the four nuts and bolts holding the flanges together (photo). Separate the flanges.

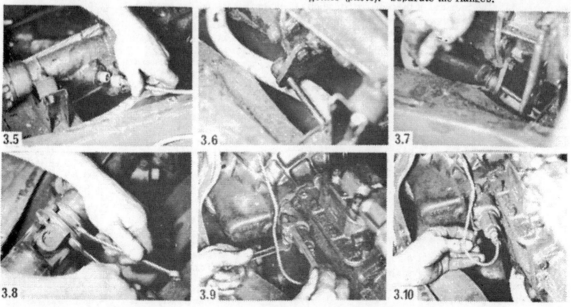

CHAPTER SIX

9. Undo the nut from the bolt which holds the slave cylinder in place (photo).

10. Remove the bolt and pull the slave cylinder complete with its pipe from off the back of the bellhousing (photo).

11. Undo the four nuts which hold the remote control extension in place and lift the extension off the top of the gearbox (photo).

12. Undo the two nuts and bolts which hold the starter motor in place and disengage the motor (photo).

13. The rear end of the gearbox extension is held to the chassis frame by two rubber mountings connected to a bracket. Undo the nut and spring washer from the stud in the centre of each mounting (photo).

14. Undo and remove all the nuts and bolts which hold the bellhousing flange in place on the rear of the engine (photo).

Fig. 6.1. EXPLODED VIEW OF THE GEARBOX

1 Gear knob.
2 Locknut.
3 Gear lever.
4 Cover.
5 Shield.
6 Plate.
7 Spring.
8 Circlip.
9 Spring.
10 Nylon ball.
11 Stepped nylon washer.
12 Bush.
13 Washer.
14 Gear lever end.
15 Reverse stop pin.
16 Locknut.
17 Bolt.
18 Plug.
19 Gasket.
20 Spring.
21 Plunger.
22 Taper locking pin.
23 1st & 2nd gear selector shaft.
24 3rd & 4th gear selector shaft.
25 Reverse gear selector shaft.
26 Interlock ball.
27 Nut.
28 Rubber 'O' ring.
29 Gearbox top cover.
30 Gasket.
31 Selector ball-end.
32 Bolt.
33 Dowel.
34 Washer.
35 Bonded rubber bush.
36 Gear change extension.
37 Reverse stop.
38 Bolt.
39 Nyloc nut.
40 Screw.
41 Pin.
42 Front remote control shaft.
43 Taper locking pin.
44 Fork.
45 Nut.
46 Rear remote control shaft.
47 Bolt.
48 1st & 2nd gear selector fork.
49 Reverse selector.
50 Interlock ball.
51 Interlock plunger.
52 3rd & 4th gear selector fork.
53 Taper locking pin.
54 Clutch housing.
55 Pin.
56 Clutch release mechanism.
57 Wedgelock bolt.
58 Plain washer.
59 Bolt.
60 Gasket.
61 Dowel.
62 Gearbox rear extension.
63 Rubber 'O' ring.
64 Peg bolt.
65 Speedometer drive gear housing.
66 Speedometer drive gear.
67 Gearbox extension ball race.
68 Oil seal.
69 Gearbox mounting rubber.
70 Mounting bracket.
71 Nut.
72 Bolt.
73 Gasket.
74 Clutch slave cylinder bracket.
75 Gearbox chain plug.
76 Speedometer driving gear.
77 Circlip.
78 Distance washer.
79 Ball race.
80 1st speed gear.
81 Spring.
82 Shim.
83 Synchromesh ball.
84 Plunger.
85 Ball.
86 2nd speed synchro hub.
87 2nd speed synchro ring.
88 Thrust washer.
89 2nd speed mainshaft gear.
90 Thrust washer.
91 Bushes.
92 3rd speed mainshaft gear.
93 Thrust washer.
94 Circlip.
95 3rd & 4th speed synchro sleeve.
96 3rd speed synchro cup.
97 3rd & 4th speed inner synchro hub.
98 4th speed synchro cup.
99 Circlip.
100 Distance washer.
101 Circlip.
102 Ball race.
103 Oil deflector.
104 Input shaft.
105 Torrington needle roller bearing.
106 Mainshaft.
107 Distance washer.
108 Driving flange.
109 Spring washer.
110 Nut.
112 Layshaft.
113 Peg bolt.
114 Spring washer.
115 Rear fixed thrust washer.
116 Rear rotating thrust washer.
117 Layshaft gear cluster.
118 Layshaft bush.
119 Front fixed thrust washer.
120 Reverse gear bush.
121 Reverse gear.
122 Reverse gear actuator.
123 Actuator pivot.
124 Plain washer.
125 Nyloc nut.
126 Reverse gear shaft.
127 Reverse shaft retaining bolt.
128 Spring washer.

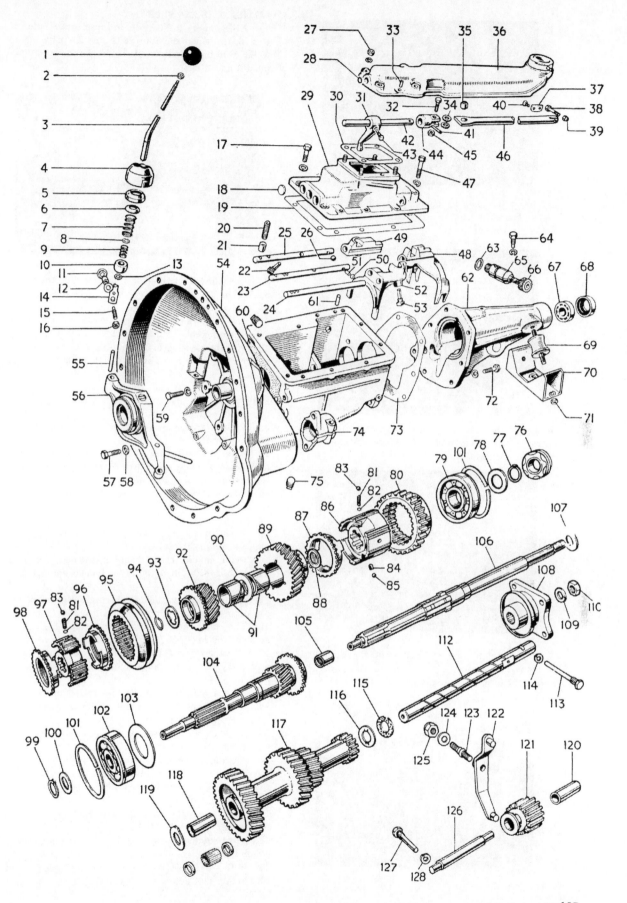

107

CHAPTER SIX

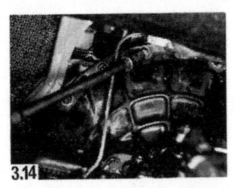

3.14

15. Raise the gearbox rear extension a few inches so the rubber mounting studs clear the slots in the mounting bracket (photo).

3.15

16. The gearbox can now be maneouvred out from inside the car (photo). **Note:** it may be necessary to support the engine to ease this task.

3.16

17. Replacement is a straightforward reversal of the removal process. Refill the gearbox with 1.5 pints of SAE 90 EP gear oil (photo).

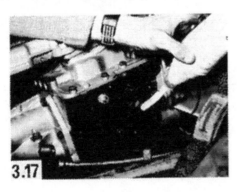

3.17

4. GEARBOX DISMANTLING
1. All numbers in brackets in this section refer to Fig. 6.1. Undo the four nuts and spring washers which hold the remote control extension in place (photo) and lift off the extension.

4.1

2. Undo the eight bolts (17, 47) which hold the gearbox cover (29) to the top of the gearbox and lift off the cover (photo).
3. The peg bolt (64) which retains the speedometer drive housing (65) is then unscrewed (photo).
4. Pull the housing (65) complete with the speedometer drive pinion out of the gearbox extension (62) photo).
5. Place the mainshaft drive flange (108) in a vice and undo the nut and spring washer (110, 109) which retains it in place (photo).
6. Pull the extension (62) away from the mainshaft flange (photo).
7. Undo the bolts (72) which retain the aluminium alloy extension to the rear of the gearbox (photo). Note the position of the longer bolt.
8. Remove the extension by tapping the underside of the mounting lug with a rawhide hammer. Lift the extension off the gearbox (photo).
9. Then lift the rear idler gear (121) from the reverse gear shaft (126).
10. From inside the bellhousing (54) undo the bolts which hold the bellhousing to the front of the gearbox (photo).
11. Separate the bellhousing from the gearbox and place the former on one side (photo).
12. Then undo the bolt and washer (113, 114) which secures the layshaft (112) in place (photo).
13. Pull the layshaft (112) out of the gearbox (photo) so the laygear drops out of mesh with the mainshaft gear.
14. With a soft metal drift carefully tap the input shaft (104) complete with bearing (102) forwards from inside the gearbox (photo).
15. As soon as the bearing is clear of the gearbox casing lift the input shaft out (photo).
16. Turning to the mainshaft (106), tap the end inside the gearbox with a rawhide or plastic headed hammer (photo) until the rear end ball bearing (79) is clear of the casing.
17. The mainshaft can now be tilted and the synchronising unit (95) slid off (photo).
18. Then remove the synchroniser ring (96) (photo), which will probably have been left in place.
19. With a pair of fine nosed circlip pliers expand the circlip (94) out of its retaining groove (photo).

20. With the aid of a couple of screwdrivers carefully ease the circlip (94) off the nose of the mainshaft (photo).

21. The mainshaft (106) can now be removed from the rear of the gearbox by driving it out with the aid of a plastic headed hammer.
22. During the final stage hold the mainshaft gear cluster together while the shaft is pulled out (photo).
23. Then lift the mainshaft gear cluster out of the gearbox and place on one side (photo).
24. Raise the laygear so it is in its normal position and measure the endfloat with a feeler gauge. The laygear (117) is then free to be lifted out (photo).
25. Remove the reverse gear shaft (126) by undoing the peg bolt (127) and spring washer (128) which holds it in place (photo).
26. Finally undo the nut (125) and bolt (123) and remove together with the operating lever (122) (photo).
27. The gearbox is now stripped right out and must be thoroughly cleaned. From the quantity of metal chips and fragments in the bottom of the gearbox worked on (photo) it is obvious that several items will be found to be badly worn. The component parts of the gearbox should now be examined for wear, and the laygear, input shaft and mainshaft assemblies broken down further as described in the following section.

5. GEARBOX EXAMINATION & RENOVATION
1. Carefully clean and then examine all the component parts for general wear, distortion, slackness of fit, and damage to machined faces and threads.
2. Examine the gearwheels for excessive wear and chipping of the teeth. Renew them as necessary. If the laygear endfloat is above the permitted tolerance of 0.0125 in. the thrust washers must be renewed. New thrust washers will almost certainly be required on any car that has completed more than 50,000 miles.
3. Examine the layshaft for signs of wear where the laygear bushes bear, and check the laygear on a new shaft for worn bushes. These are simply drifted out if new bushes are to be fitted.
4. The three synchroniser rings (87, 96, 98) are bound to be badly worn and it is a false economy not to renew them. New rings will improve the smoothness and speed of the gearchange considerably.
5. The needle roller bearing and cage (105) located between the nose of the mainshaft and the annulus in the rear of the input shaft is also liable to wear, and should be renewed as a matter of course.
6. Examine the condition of the three ball bearing assemblies, one on the input shaft (102), one on the mainshaft (79) and the other in the tail of the gearbox extension (67). Check them for noisy operation, looseness between the inner and outer races, and for general wear. Normally they should be renewed on a gearbox that is being rebuilt.
7. Examine the mainshaft bushes and fit them on the mainshaft to check for overall endfloat.
8. Fit the inner thrust washer (88) onto the mainshaft, then one of the bushes (91), the washer (90), the remaining bush (91), the thrust washer (93), and finally the circlip. With a feeler gauge measure the endfloat between the inner thrust washer and the adjacent bush. This should be between 0.004 and 0.010 in. If outside these figures experiment with alternative thrust washers until the endfloat is correct.
9. To dismantle the synchromesh units, first wrap a length of clean rag completely round a unit and then pull off the outer synchro slave. The cloth will catch the spring loaded balls and springs which are bound to fly out. Compare the length of the old springs with new and replace any that are worn. Note that an interlock plunger and ball is fitted to the second speed synchromesh hub.
10. The remote control gearchange is bound to be worn but this is dealt with in Section 10.

6. INPUT SHAFT - DISMANTLING & REASSEMBLY
1. Place the input shaft in a vice splined end uppermost and, with a pair of circlip pliers, remove the circlip which retains the ball bearing in place (photo), and then take off the distance washer.
2. Slightly close the vice with the bearing resting on top of the jaws. Tap the shaft through the bearing with a soft headed hammer (photo) and remove the oil deflecter plate.
3. Prise out the old bearing from the annulus and fit a new roller bearing assembly in place (photo).
4. With the aid of a socket spanner of slightly smaller diameter than the annulus tap the roller bearing into place (photo).
5. Fit the circlip over a new ball bearing so the concave side of the clip faces the narrowest portion of the bearing rim (photo).
6. With the aid of a block of wood and the vice tap the bearing into place on the shaft as shown.
7. Finally refit the distance washer and the circlip (photo).

7. MAINSHAFT - DISMANTLING & REASSEMBLY
1. The mainshaft has to be partially dismantled before it is possible to remove it from the gearbox (Section 4 paras. 16 to 23). Final mainshaft dismantling consists of removing the nylon speedometer driver gear and the ball bearing.
2. To remove the nylon speedometer drive gear, select an open ended spanner which just fits over the mainshaft and then lay the spanner across the jaws of the vice under the drive gear. Carefully tap the mainshaft downwards with a soft headed hammer, so driving off the drive gear (photo).
3. Remove the bearing retaining circlip (photo) and distance washer and tap the shaft out of the bearing as described previously.
4. Fit a new bearing onto the mainshaft and with the aid of the vice tap the new bearing into place (photo).

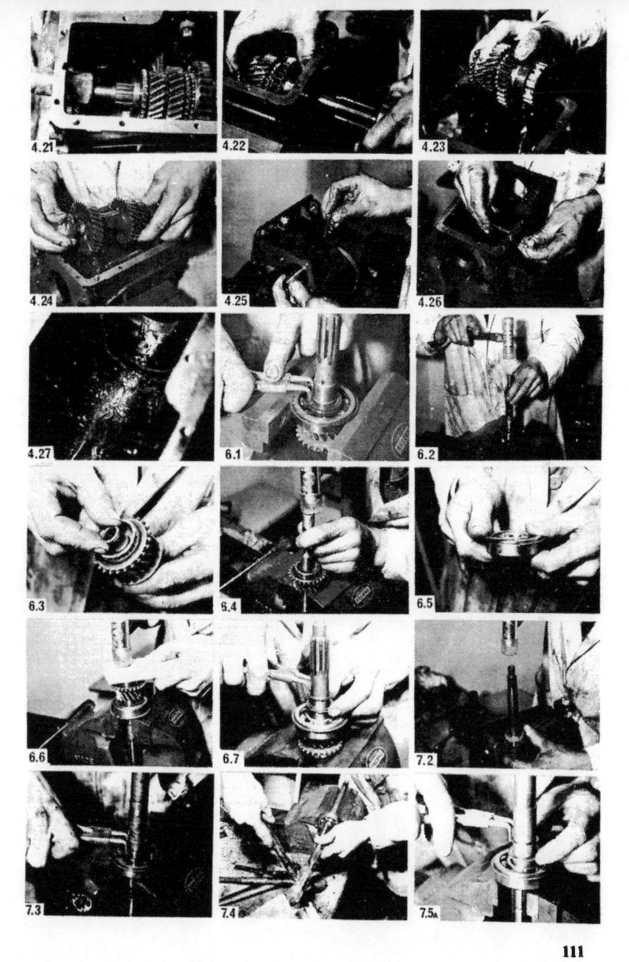

5. Then fit the distance washer and the retaining circlip (photo). Finally tap the nylon speedometer drive gear into place using the side of the jaws of the vice as a press (photo).

8. GEARBOX REASSEMBLY

1. Refit the gear selector lever to the side of the gearbox and replace the securing nut and plain washer (photo).
2. Smear the laygear endface adjacent to the small straightcut gear with thick grease and fit the special thrust washer (photo).
3. Fit a new laygear thrust washer to each end of the gearbox with the aid of thick grease so the bronze faces are adjacent to the laygear and the thrust washer tags rest in the recesses in the casing (photo).
4. Carefully lower the laygear into place taking care not to disturb any of the washers (photo). Temporarily fit the layshaft and measure the laygear endfloat with a feeler gauge. Endfloat should be between 0.0125 in. and 0.006 in. Rub down the backs of the thrust washers if more clearance is required. DO NOT rub down the bronze portion adjacent to the laygear. Remove the layshaft.

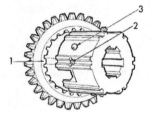

Fig. 6.2. Second speed synchro unit, showing: 1 'Motor' spline. 2 Interlock bolt. 3 and the synchro ball.

5. Reassemble a new synchroniser ring to the second gear synchroniser hub making sure that the three lugs on the ring locate in the cut outs in the hub. Then fit the rear thrust washer (88) with its scrolled face upwards and fit second gear and bush inside the synchroniser ring (photo).
6. Next fit the centre thrust washer and third gear and bush (photo) and finally the front thrust washer (93) with its scrolled face down.
7. Pass the end of the mainshaft into the gearbox and fit the assembled gearcluster over the end of the shaft (photo).
8. With the aid of a screwdriver carefully work the circlip into place (photo). During this operation press each side of the circlip in turn to keep it square on the mainshaft.
9. Then fit a new synchroniser ring with the three lugs facing forwards (photo).
10. Fit the third and top synchroniser hub to the mainshaft with the longer boss of the inner synchro member facing forwards (photo).
11. Screw a nut over the end of the mainshaft to protect the threads and carefully drive the mainshaft assembly complete with the rear ball bearing race into place.
12. Fit a new synchroniser ring to the front of the third and top synchro hub so the tabs in the former engage with the slots in the latter and then fit the input shaft in place (photo).
13. Tap the shaft and bearing fully home with the aid of a soft faced hammer (photo).
14. To align the thrust washers and the laygear push a tapering rod through the gearbox and laygear (photo).
15. With the thrust washers and laygear correctly positioned feed in the layshaft oiling it as it enters the gearbox (photo). Ensure the lockpin hole enters the casing last.
16. Align the lockpin hole in the shaft with the hole in the casing and fit the lockpin and washer (photo).
17. Then fit the reverse gear shaft into the casing and align its locating hole with the hole in the casing and insert the lockpin and washer (photo).
18. Carefully engage the pin on the lower arm of the operating lever, with the groove cut on one side of the reverse gear (photo).
19. Finally push the reverse gear fully home (photo). TAKE GREAT CARE that reverse gear does not move forwards and disengage with the operating lever when the rear extension is being fitted.
20. If the gearbox is being rebuilt it is a false economy not to renew the rear oil seal and ball race. Mount the rear extension in a vice and with the aid of a drift drive out the oil seal and ball race from inside the extension (photo).
21. Carefully tap a new bearing into place ensuring it is square in the bore (photo).
22. Then tap the oil seal into place (photo) with the sealing lip facing forward.

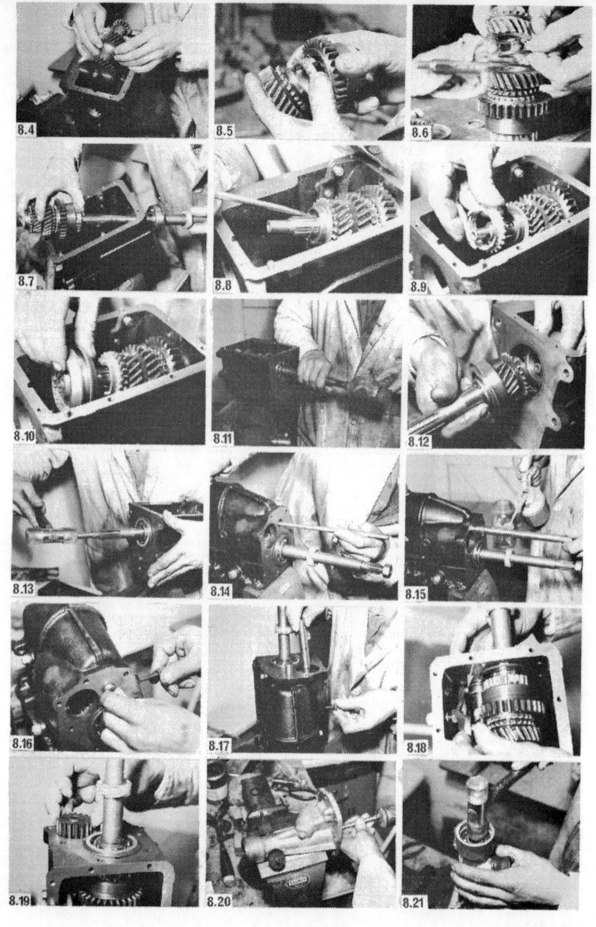

CHAPTER SIX

23. Fit the special distance washer (107) in place on the end of the mainshaft (photo).

24. Then fit a new gasket to the rear face of the gearbox casing using gasket cement if wished (photo).
25. Carefully lower the rear extension into position (photo) tapping it home with a soft headed hammer if need be.
26. Replace the bolts and washers securing the extension in place not omitting the special bracket (photo).
27. Then refit the mainshaft drive flange and do up the nut and washer which holds it in place (photo).
28. Refit the speedometer drive gear assembly, making sure the rubber washer is in place in the groove, and that the hole for the lock bolt and washer is in line with the hole in the casing (photo).
29. Fit a new gasket to the bellhousing/gearbox flange. If a new gasket is not to hand cut a new gasket from stiff brown paper. Lay the paper over the bellhousing endface, and holding the paper taut make a series of rapid gentle taps with a ball headed hammer. This will soon cut the paper to the desired shape (photo). Proper gasket paper is even better than brown paper.
30. Offer up the bellhousing to the gearbox (photo) and replace the securing bolts and washers.
31. Finally, ensuring that all the selectors and gears are in neutral, fit the top cover and remote control extension in place using new gaskets.

9. GEAR SELECTORS - REMOVAL & REPLACEMENT

1. Position the selector shafts (25, 23, 24) so that they are as far forward as possible and then drive out the welch plugs (28) with a $1/8$ in. punch positioned in turn through the small holes just inside the end of the cover (photo).
2. Undo the threaded tapered locking bolts (53, 22) from the selector shafts and forks.
3. Push the reverse gear selector shaft (25) out, followed by the other two (23, 24). The two interlock balls (26, 50), plunger (51), three selector plungers (21) and springs (20) can then be removed. One spring (20) and one plunger (21) will emerge from each of the three holes indicated in the photo.
4. Reassembly commences by fitting the springs and plungers in place and then sliding the third and top selector shaft (24) through the third and top selector fork in the top cover. Press down the selector plunger to allow the shaft to pass over it, and continue pushing the shaft home until it is in the neutral position, i.e., the plunger is resting in the centre one of the three cut outs on the shaft.
5. Replace the reverse gear selector shaft (25) and selector fork (49) in the same way ensuring it too is in neutral.
6. Then fit the interlock plunger (51) to the first and second gear selector shaft (23), and slide the shaft into place in the selector fork (48), noting that the shaft also passes through the third and top selector fork (24). Before the shaft (23) is fully home, drop the two interlock balls (50, 26) in through the centre selector shaft hole so that one ball seats each side of the transverse bore which connects the selector shaft bores. The centre selector shaft can then be pushed further in until the plunger is resting in the centre of the three cut outs, and the interlock balls and plunger are held by the shafts.
7. Refit the threaded tapered lock bolts and refit the welch plugs using sealing compound to give a leakproof joint.

10. REMOTE CONTROL ASSEMBLY - OVERHAUL

1. Certain items in the remote control gearchange are prone to wear and these should always be renewed when the gearbox is being overhauled. The items concerned are small and relatively inexpensive and comprise the plate (6), reverse gear spring (7), a smaller spring (9), the nylon ball (10), the bonded rubber bush (35) and washers (34), and the bush (12) and washers (11). (All numbers refer to Fig. 6.1.)
2. To renew the items on the gearlever press down and twist off the cover (4) and remove the shield plate and larger spring (photo).
3. Undo and remove the nut and bolt which holds the gearlever to the rear remote control shaft (photo).
4. Lift out the gearlever and place it in a vice so the circlip which retains the small spring in place can be removed, and the spring and nylon ball removed (photo).
5. Fit a new nylon ball and spring and with the aid of a spanner which just fits over the gearlever tap the circlip home until it rests in its groove.
6. Then fit a new bush and washer to the end of the gearlever (photo).
7. Undo the nut and bolt which holds the rear remote control rod to the fork (photo).
8. As can be seen in the photograph the old bush (bottom) had completely broken up. Press out the remains of the old bush.
9. This is most easily done in a vice using two sockets, one considerably larger, and one fractionally smaller than the bush. Place the sockets either side of the shaft and use the small socket to push the bush

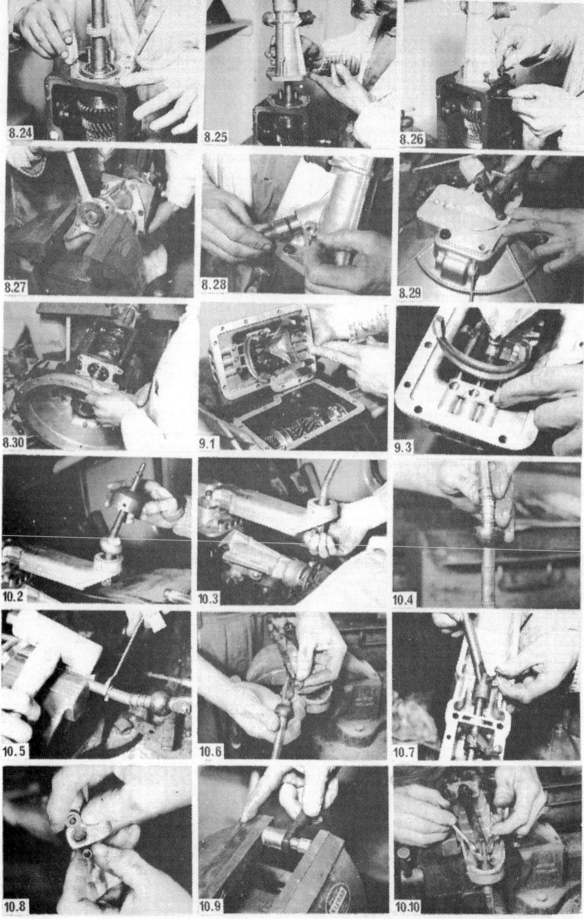

CHAPTER SIX

into the larger socket (photo).

10. Carefully press the new bush into place and reconnect the rod to the shaft using new washers. Fit the gearlever to the extension and refit the nut and bolt which secures the gearchange lever to the remote control shaft (photo).

11. Then refit the larger spring and the remaining components to complete the assembly (photo).

FAULT FINDING CHART

Cause	Trouble	Remedy
SYMPTOM:	**WEAK OR INEFFECTIVE SYNCHROMESH**	
General wear	Synchronising cones worn, split or damaged.	Dismantle and overhaul gearbox. Fit new gear wheels and synchronising cones.
	Baulk ring synchromesh dogs worn, or damaged	Dismantle and overhaul gearbox. Fit new baulk ring synchromesh.
SYMPTOM:	**JUMPS OUT OF GEAR**	
General wear or damage	Broken gearchange fork rod spring	Dismantle and replace spring.
	Gearbox coupling dogs badly worn	Dismantle gearbox. Fit new coupling dogs.
	Selector fork rod groove badly worn	Fit new selector fork rod.
	Selector fork rod securing screw and locknut loose	Remove side cover, tighten securing screw and locknut.
SYMPTOM:	**EXCESSIVE NOISE**	
Lack of maintenance	Incorrect grade of oil in gearbox or oil level too low	Drain, refill, or top up gearbox with correct grade of oil.
General wear	Bush or needle roller bearings worn or damaged	Dismantle and overhaul gearbox. Renew bearings.
	Gearteeth excessively worn or damaged	Dismantle, overhaul gearbox. Renew gearwheels.
	Laygear thrust washers worn allowing excessive end play	Dismantle and overhaul gearbox. Renew thrust washers.
SYMPTOM:	**EXCESSIVE DIFFICULTY IN ENGAGING GEAR**	
Clutch not fully disengaging	Clutch pedal adjustment incorrect	Adjust clutch pedal correctly.

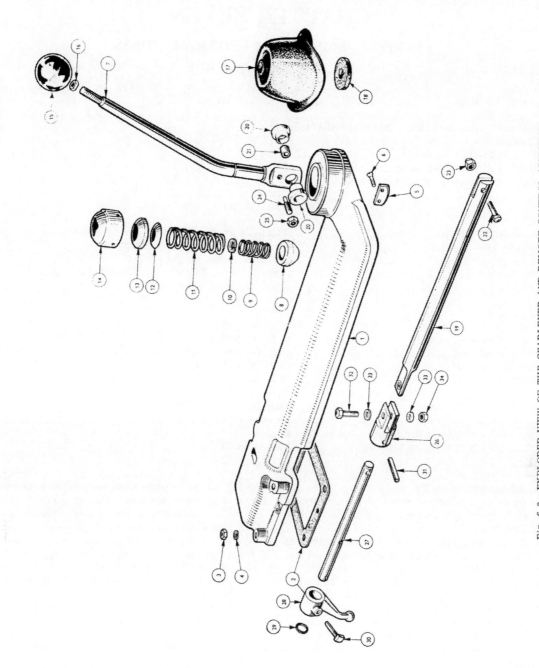

Fig. 6.3. EXPLODED VIEW OF THE GEARLEVER AND REMOTE CONTROL ASSEMBLY

1 Remote control extension. 2 Gasket. 3 Nut. 4 Lock washer. 5 Reverse baulk plate. 6 Screw. 7 Gear lever. 8 Spherical bush. 9 Reverse baulk spring. 10 Circlip. 11 **Spring.** 12 Dished inner washer. 13 Dished outer washer. 14 Cap. 15 Gearlever knob. 16 Nut. 17 Boot. 18 Rubber washer. 19 Operating shaft. 20 Nylon bush. 21 Sleeve. 22 **Bolt.** 23 Nyloc nut. 24 Screw. 25 Locknut. 26 Coupling. 27 Gearlever shaft. 28 Internal gear lever. 29 Oil seal. 30 Wedglok taper set bolt. 31 Dowel. 32 Bolt. 33 Plain washer. 34 Nyloc nut.

117

CHAPTER SEVEN

PROPELLER SHAFT AND UNIVERSAL JOINTS

CONTENTS

General Description...	1
Propeller Shaft - Removal & Replacement	2
Drive Shaft Universal Joints - Removal & Replacement..	3
Universal Joints - Inspection & Repair	4
Universal Joints - Dismantling..	5
Universal Joints - Reassembly..	6
Strap Drive - Dismantling & Reassembly..	7

SPECIFICATIONS

Propeller Shaft - Type	BRD solid Pt. No. 207410
Alternative	Hardy Spicer solid Pt. No. 208033
Alternative	BRD frictionless Pt. No. 209834
Alternative	Hardy Spicer sliding joint Pt. No. 211143
Alternative	BRD strap drive Pt. No. 212549
Length of Shafts	
BRD solid	50.25 in. (127.6 cm.)
Hardy Spicer solid..	50.13 in. (127.3 cm.)
BRD frictionless	49.9 in. (126.8 cm.)
Hardy Spicer sliding joint	50.2 in. (127.5 cm.)
BRD strap drive	50.2 in. (127.5 cm.)
Universal Joint Lubricant..	Shell Dentax 250 or Retinax A
U.J. splines lubricant	Duckhams grease Q5648
No. of U.J's on propeller shaft	2
No. of U.J's on each drive shaft	1 (total No. 2)

1. GENERAL DESCRIPTION

Drive is transmitted from the gearbox to the differential unit by means of a finely balanced propeller shaft.

Fitted at each end of the shaft is a universal joint which allows for slight movement of the engine, gearbox and differential units in their mountings. A further universal joint is fitted to each of the differential unit drive flanges to accommodate the movements of the swing axle rear suspension.

Each universal joint comprises a four legged centre spider, four needle roller bearings and cups, two yokes, and the necessary seals, retainer, and circlips.

The propeller shaft and universal joints are fairly simple components and to overhaul and repair them is not difficult. It should be noted that when the propeller shaft universal joints become worn and develop radial play, to obtain the best result always fit a complete exchange propeller shaft which will already be balanced to the fine degree necessary for smooth running. Overhauling the universal joints on the existing shaft may or may not throw the shaft out of balance. The universal joints are all of the sealed type and require no maintenance. The threaded holes normally filled by grease nipples are fitted with plugs. These plugs can be removed and grease nipples fitted for lubrication purposes if wished, but note that some of the plugs have left-hand threads and they are very difficult to get at with the shaft in situ. Do not run the car on the road with grease nipples fitted instead of plugs as there is a possibility of fouling.

2. PROPELLER SHAFT - REMOVAL & REPLACEMENT

1. Jack up the rear of the car, or position the rear of the car over a pit or on a ramp.
2. If the rear of the car is jacked up supplement the jack with support blocks so that danger is minimised should the jack collapse.
3. If the rear wheels are off the ground place the car in gear or put the handbrake on to ensure that the propeller shaft does not turn when an attempt is

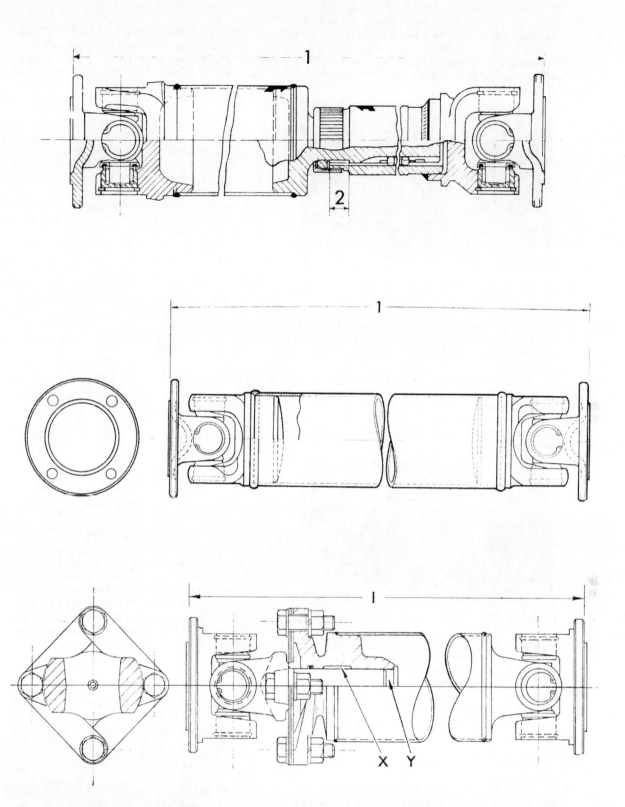

Fig. 7.1. THREE TYPES OF PROPELLER SHAFT
Top - Frictionless propeller shaft.   Centre - Solid propeller shaft.   Bottom - Strap driven propeller shaft.

made to loosen the four nuts on each flange.

4. Take off the gearlever knob and grommet, remove the fascia support where fitted, and undo the fasteners which hold the gearbox cover to the bulkhead. NOTE that there are three screws which are removed from the engine side of the bulkhead. Lift off the gearbox cover to expose the gearbox and front U.J. bolts.

5. The propeller shaft is carefully balanced to fine limits and it is important that it is replaced in exactly the same position it was in prior to its removal. Scratch a mark on both the propeller shaft flanges and the gearbox/rear axle drive flanges to ensure accurate mating when the time comes for reassembly.

6. Unscrew and remove the four self-locking nuts, bolts, and securing washers which hold each flange on the propeller shaft to the flanges on the rear axle and gearbox (photo). Lever the engine/gearbox unit forward slightly to free the propeller shaft and pull it out.

7. Replacement of the propeller shaft is a reversal of the above procedure. Ensure that the mating marks scratched on the sides of the propeller shaft flanges line up with those on the rear axle and gearbox flanges, and always use new nyloc nuts. Note that if the sliding joint (or strap drive) type of propeller shaft is fitted, this end must be adjacent to the differential unit.

3. DRIVE SHAFT UNIVERSAL JOINTS - REMOVAL & REPLACEMENT

1. If the drive shaft universal joints are worn the drive shafts must be removed from the car before the universal joints can be dismantled.

2. Removal and replacement of the drive shafts is covered in Chapter 8, Section 4. Once the drive shafts are out the universal joints can be dealt with in an identical manner to those for the propeller shaft.

4. UNIVERSAL JOINTS - INSPECTION & REPAIR

1. Wear in the needle roller bearings is characterised by vibration in the transmission, 'clonks' on taking up the drive, and in extreme cases of lack of lubrication, metallic squeaking, and ultimately grating and shrieking sounds as the bearings break up.

2. It is easy to check if the needle roller bearings are worn with the propeller shaft in position, by trying to turn the shaft with one hand, the other hand holding the rear axle flange when the rear universal is being checked, and the front gearbox coupling when the front universal is being checked. Any movement between the propeller shaft and the front and the rear half couplings is indicative of considerable wear.

3. If worn, the old bearings and spiders will have to be discarded and a repair kit, comprising new universal joint spiders, bearings, oil seals, and retainers purchased. Check also by trying to lift the shaft and noticing any movement in the joints.

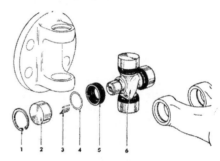

Fig. 7.2. EXPLODED VIEW OF A UNIVERSAL JOINT
1 Circlip. 2 Cup. 3 Needle rollers. 4 Washer. 5 Seal. 6 Spider.

5. UNIVERSAL JOINTS - DISMANTLING

1. Clean away all traces of dirt and grease from the circlips located on the ends of the bearing cups, and remove the clips by pressing their open ends together with a pair of pliers (photo), and lever them out with a screwdriver. NOTE: If they are difficult to remove tap the bearing cup face resting on top of the spider with a mallet which will ease the pressure on the circlip.

PROPELLER SHAFT AND UNIVERSAL JOINTS

2. Take off the bearing cups on the propeller shaft yoke. To do this select two sockets from a socket spanner set, one large enough to fit completely over the bearing cup (15/16 AF or slightly larger) and the other smaller than the bearing cup (photo), i.e. 5/8 AF or slightly less.

3. Open the jaws of the vice and with the sockets opposite each other and the U.J. in between tighten the vice and so force the narrower socket to move the opposite cup partially out of the yoke (photo) into the larger socket.

4. Remove the cup with a pair of pliers (photo). Remove the opposite cup, and then free the yoke from the propeller shaft.

5. To remove the remaining two cups now repeat the instructions in paragraph 3, or use a socket and hammer as illustrated.

6. UNIVERSAL JOINTS - REASSEMBLY
1. Thoroughly clean out the yokes and journals. Smear jointing compound on the journal shoulders of the new spider if a retainer is used. With the aid of a tubular drift fit the oil seal retainers to the trunnions and then fit the oil seals to the retainers.
2. On some models it will be found that there is no retainer, merely a rubber seal. NOTE: The spider must always be fitted so the lubricating plug holes are towards the propeller shaft.
3. On replacement great care must be taken to ensure the journals and associated parts are absolutely clean. Fill the grease holes in the spider journal with the recommended lubricant (see specifications) making sure all air bubbles are eliminated. Fill each bearing assembly to a depth of approximately 1/8 in. (3 mm.)
4. Fit new rubber seals to the spiders and then replace the spiders and bearings in the yokes. Refit the circlips.

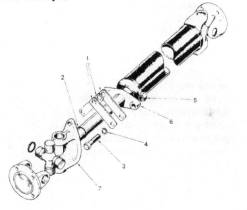

Fig. 7.3. EXPLODED VIEW OF THE STRAP DRIVE PROPELLER SHAFT
1 Straps. 2 Shaft. 3 Bolt. 4 Washer. 5 Nylon nut. 6 Tube coupling. 7 End yoke.

7. STRAP DRIVE - DISMANTLING & REASSEMBLY
1. Referring to Fig. 7.3. undo the four bolts (3) and the nuts and washers which hold the connector straps (1) to the tube coupling (6) and end yoke (7) The end yoke can then be pulled clear of the tube coupling.
2. On reassembly pack the bore at 'Y' in Fig. 7.1. with Duckham Q.5648 grease or equivalent and generously oil the area 'X' with Shell Dentax 250, Retinax 'A' or equivalent. Ensure the straps are interleaved as shown in Fig. 7.3.

CHAPTER EIGHT

REAR AXLE

CONTENTS

General Description...	1
Routine Maintenance...	2
Differential Unit - Removal & Replacement...	3
Rear Drive Shafts - Removal & Replacement...	4
Rear Drive Shaft Hubs - Removal, Dismantling & Reassembly...	5
Pinion Oil Seal - Removal & Replacement...	6
Differential Carrier Assembly - Removal & Replacement...	7
Differential Inner Drive Shafts - Removal & Replacement...	8

SPECIFICATIONS

Type...	Hypoid bevel gears
Ratios: 948 c.c. Herald Saloons...	4.875 to 1
: 948 c.c. Herald Coupes...	4.55 to 1
: Herald 1200, 12/50 & 13/60...	4.11 to 1
Differential bearing preload...	.003 in.
Pinion bearing preload...	12 to 16 lb/in.
Backlash adjustment - crown wheel...	Shims
Backlash adjustment - pinion...	Shims
Rear axle oil capacity...	1.0 pints (1.2 U.S. pints, .70 litres)
Drive shafts...	Independent swing type
Rear axle recommended lubricant...	S.A.E. Hypoid 90 gear oil

TORQUE WRENCH SETTINGS

Bearing cap to housing...	32 to 34 lb/ft. (4.425 to 4.70 kg/m.)
Crown wheel to differential casing...	40 to 45 lb/ft. (5.530 to 6.221 kg/m.)
Hypoid housing...	16 to 18 lb/ft. (2.212 to 2.489 kg/m.)
Hypoid pinion flange attachment...	70 to 85 lb/ft. (9.678 to 11.752 kg/m.)
Mounting plate to hypoid housing...	26 to 28 lb/ft. (3.595 to 3.871 kg/m.)
Rear axle mounting plate to frame...	26 to 28 lb/ft. (3.595 to 3.871 kg/m.)
Rear axle to frame...	38 to 40 lb/ft. (5.254 to 5.530 kg/m.)
Hub to axle shaft...	100 to 110 lb/ft. (13.826 to 15.21 kg/m.)
Shaft joint to inner axle shaft...	24 to 28 lb/ft. (3.318 to 3.595 kg/m.)

GENERAL DESCRIPTION

The main rear axle component is the hypoid differential unit which is fixed to the chassis at four points. Swing axle drive shafts, pivoting at their inner ends on universal joints attached to the differential drive flanges, carry the drive to the hubs via needle roller and ball bearings carried in trunnions mounted in the centre of vertical links. The upper ends of the links are attached to a single transverse leaf spring and the lower ends to longitudinal tie rods. The tie rods make use of flexible bushes at each end and provide fore and aft location for the links.

The crown wheel and pinion each run on opposed taper roller bearings, the bearing preload and meshing of the crown wheel and pinion being controlled by shims. Spring loaded oil seals, of the type normally found at the front of the differential nose piece, prevent oil loss from the differential at the pinion and drive shaft holes.

2. ROUTINE MAINTENANCE

1. Every 6,000 miles remove the filler plug on the side of the differential unit and if no oil seeps out, top up with Castrol hypoy or a similar E.P. 90 hypoid gear oil. If more than a very small amount is required it is likely that the oil is leaking. Investigate and rectify. NOTE: On post 1965 models it is recommended that the oil is checked just once every 12,000 miles.

2. Every 12,000 miles (24,000 miles on post 1965 models) drain the oil when hot, clean the drain plug,

Rear Axle

Fig. 8.1. 1 Oil level plug. 2 Oil drain plug.

and refill the differential unit with one pint of hypoid S.A.E. 90 gear oil.

3. DIFFERENTIAL UNIT - REMOVAL & REPLACEMENT

1. Remove the rear wheel hub caps and loosen the wheel nuts one turn.
2. Jack up the rear of the car and place stands or other supports under the chassis. For safety's sake disconnect the battery.
3. Drain the oil from the differential unit and then remove both rear wheels placing the wheel nuts in the hub caps for safe keeping.
4. Place jacks under the vertical link on each side of the car as shown in Fig. 8.2. and raise the jack slightly to take the spring load. Take off the nyloc nuts and washers from the damper lower attachment eyes and then pull the bottom of the dampers clear of the mounting pins. Lower the jacks.

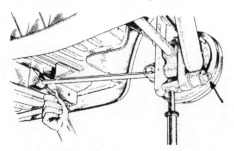

Fig. 8.2. Supporting the vertical links with a screw jack. To free the bottom of the damper undo the nut arrowed.

5. Remove the rear exhaust pipe and silencer by undoing the pipe clip at the rear of the expansion chamber, free the silent block mounting from the hypoid front plate and take out the nut and bolt which holds the fabric strap in place.
6. Mark the adjacent differential to propeller shaft and drive shaft flanges and undo and remove the four bolts and nyloc nuts from these three sets of couplings.
7. From behind the centre of the front seats undo the four attachments which hold the spring access cover in place and lift the cover away.
8. Access can now be gained to the nyloc nuts which hold the spring securing plate in place. Undo the nuts and remove the plate. Unscrew the studs from the top of the differential unit.
9. Take the weight of the hypoid unit on a jack, or better still get a friend to hold it, undo the nyloc nuts, and remove the large plain washers and rubber bushes which hold each end of the front mounting plate to the studs on the chassis. Undo and remove the two nuts and bolts which hold the differential casing lugs to the chassis brackets and carefully manoeuvre the differential unit out from under the car. (See Fig. 8.8.).
10. Replacement commences by refitting the differential unit lugs to the rear mounting points on the chassis, doing up the nuts on the retaining bolts finger tight. Refit the rubber pads to the ends of the front mounting plate taking great care to fit the rubbers the right way up (see Fig. 8.8.), and renew any of the rubbers that are worn or perished. Replace the plain washers and tighten down all the nyloc nuts securely.
11. Then refit the spring studs, and spring plate, followed by the drive shaft and propeller shaft couplings, the dampers, exhaust tail pipe and silencer and wheels.
12. Fill the hypoid unit with one pint of S.A.E. 90 hypoid gear oil, and adjust the brakes if necessary.

4. REAR DRIVE SHAFTS - REMOVAL & REPLACEMENT

1. The drive shafts are removed complete with the hubs and vertical links. It is necessary to remove the shafts if attention is required to either the universal joints or the hub bearings.
2. Working on one side of the car, remove the hub caps, and loosen the wheel nuts. Jack up the rear of the car and support it on stands. Take off the road wheel.
3. Remove the brake master cylinder filler cap and then refit the cap over a piece of cellophane. Undo the brake metal pipe union nut (3 in Fig. 8.3.), and then holding the hexagon on the brake hose undo the hose attachment nut (2) and washer.

Fig. 8.3. 1 Flexible hose. 2 Locknut. 3 Union nut. 9 Eye bolt nut.

4. Free the clevis pin from the lever on the backplate which is attached to the handbrake cable lever.
5. Place a jack under the vertical links as shown in Fig. 8.2., and screw the jack up slightly to relieve the mounting pin and nut (arrowed) which secures the lower end of the damper, of all load. Remove the nut and pull the damper off the pin.

123

CHAPTER EIGHT

6. Undo the bolt which holds the radius arm to its bracket on the vertical link and undo the four bolts and nuts which hold the differential drive flange to the flange on the universal joint after having made a mating mark across the flanges to ensure identical positioning on reassembly.

7. Remove the jack, undo the nut which holds the spring eye bolt in place, and supporting the brake drum with one hand pull out the spring eye bolt as shown in Fig. 8.4.

Fig. 8.4. Pulling out the spring eye bolt.

8. The drive shaft can now be removed complete with the hub and vertical link.

9. Replacement commences by refitting the vertical link to the spring, tightening the nyloc nut on the spring eye bolt finger tight only at this stage.

10. Jack up the vertical link so the damper can be refitted and reconnect the universal joint to the differential flange and the radius arm to the bracket on the front of the vertical link.

11. Reconnect and bleed the brakes, refit the road wheel, lower the car to the ground and with one person sitting in the car to give a 'static laden' condition tighten the nyloc nut in the spring eye bolt securely.

5. REAR DRIVE SHAFT HUBS - DISMANTLING, EXAMINATION & REASSEMBLY

1. It is virtually impossible to dismantle the hub without using a hub extractor and it may be necessary to take the assembly to a garage with the necessary Churchill tool No. S109C.

2. Place the drive shaft between the protected jaws of a vice, and take off the brake drum after undoing the countersunk screws and backing off the brake shoe adjustment.

3. Referring to Fig. 8.5. undo the hub nut (31) and washer and then pull off the hub using a hub extractor. Note the key (42).

4. Bend back the lock tabs (71) and undo the four bolts (70) which hold the grease trap (32) brake backplate (72) outer seal housing (33) oil seal (34) and gasket (36) to the trunnion housing (37).

5. To free the vertical link (75) from the trunnions (37) undo the nyloc nut (74) and then wriggle out the bolt (73).

6. Now push the trunnion (37) further onto the drive shaft so that the ball race (35) protrudes enough for the fitting of an extractor, the one to use being

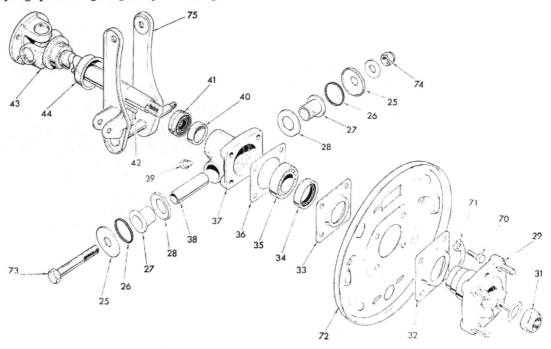

Fig. 8.5. EXPLODED VIEW OF THE OUTER DRIVE SHAFT AND HUB ASSEMBLY

25 Large flanged washers. 26 Rubber seals. 27 Flanged nylon bush. 28 Flanged washer. 29 Hub. 31 Nut. 32 Grease trap. 33 Seal housing. 34 Oil seal. 35 Ball bearing. 36 Gasket. 37 Trunnion. 38 Steel sleeve. 39 Grease plug. 40 Needle roller bearing. 41 Oil seal. 42 Key. 43 Universal joint. 44 Oil seal plunger. 70 Bolt. 71 Tab washer. 72 Brake backplate. 73 Bolt. 74 Nyloc nut. 75 Vertical link.

REAR AXLE

Churchill tool No. S4221A. With the ball race bearing removed, the shaft can be pulled out of the trunnion.

7. Carefully remove the oil seal (41) and then with the trunnion placed on a block of wood with its flanged face downwards drift out the needle roller bearing (40) with the aid of a suitable drift or Churchill tool No. S300. (Fig. 8.5. refers).

8. Thoroughly clean all the parts and renew the bearings if they are worn, rough or chipped. Examine the hub for cracks, a worn taper and keyway, and worn outer oil seal contacting faces. Check the shaft for straightness and wear. If necessary get expert advice from your local Triumph distributor, as wear that appears to be very small can have a considerable effect on performance.

9. Reassembly commences by fitting the needle roller bearing radiused end first (i.e. pressing on the lettered end) into the trunnion with the aid of Churchill tool No. S300, and tool No. S4221A which is now used as a press until the bearing is 0.5 in. (12.7 mm.) from the small end of the trunnion face.

10. Drift the inner oil seal into the trunnion housing with the sealing lips trailing and then drive the oil plunger (44) onto the axle shaft.

11. Thoroughly grease the needle rollers and slide the drive shaft through the trunnion taking great care that the lips of the seal are not turned back or damaged.

12. Fit the drive shaft between the protected jaws of a vice and then drift on the ball race (35) which should be prepacked with grease. Fit a new oil seal (34) ensuring that the sealing lip is trailing.

13. Fit a new gasket (36) to the trunnion flange face (37) and then fit the seal housing (33), brake backplate (72) with the wheel cylinder uppermost, and the grease trap (32) with the duct facing downwards.

14. Then fit the four securing bolts (70) and tab washers (71). Check that the key (42) fits tightly in the slot in the drive shaft and also the hub. Drive the hub (29) on squarely, replace the washer and nut (31) and tighten to 110 lb.ft. Refit the brake drum and replace the countersunk screws.

15. Finally reassemble the vertical links to the trunnion housing. Ensure that the various bushes, rubber seals and steel sleeve are in the order shown in Fig. 8.5. Refit the bolt (73) and tighten the nut (74) to a torque of 40 lb.ft.

6. PINION OIL SEAL - REMOVAL & REPLACEMENT

If oil is leaking from the front of the differential nose piece it will be necessary to renew the pinion oil seal. If a pit is not available, jack and chock up the rear of the car. It is much easier to do this job over a pit, or with the car on a ramp. First remove the exhaust tail pipe.

1. Mark the propeller shaft and pinion drive flanges to ensure their replacement in the same relative positions.

2. Unscrew the nuts from the four bolts holding the flanges together remove the bolts and separate the flanges.

3. If the oil seal is being renewed with the differential nose piece in position, drain the oil and check that the handbrake is firmly on to prevent the pinion flange moving.

4. Pull out the split pin and unscrew the nut in the centre of the pinion drive flange. Although it is tightened down to a torque of 85 lb/ft. It can be removed fairly easily with a long extension arm fitted to the appropriate socket spanner. Remove the nut and spring washer.

5. Pull off the splined drive flange, which may be a little stubborn, in which case it should be tapped with a hide mallet from the rear; the pressed steel end cover; and prise out the oil seal with a screwdriver taking care not to damage the lip of its seating.

6. Replacement is a reversal of the above procedure. NOTE that the new seal must be pushed into the differential nose piece with the edge of the sealing ring facing inwards, and take great care not to damage the edge of the oil seal when replacing the end cover and drive flange. Smear the face of the flange which bears against the oil seal lightly with oil before driving the flange onto its splines. Tighten the nut to 85 lb/ft.

7. DIFFERENTIAL CARRIER ASSEMBLY - REMOVAL & REPLACEMENT

1. If it is wished to renew the differential carrier assembly remove the differential unit from the car as described in Section 3.

2. Thoroughly clean the unit with paraffin and then remove the short drive shafts as described in Section 8.

3. Undo the ring of bolts (63) and spring washers which hold the front casing and differential carrier assembly to the rear casing. (Fig. 8.8, page 127).

4. Turn the pinion so that the two chamfered parts on the edge of the differential carrier allow removal of the differential housing.

5. On replacement thoroughly clean the differential unit joint faces and fit a new paper joint. Note specially that the two short flange bolts must be fitted next to the short drive shaft.

6. Further dismantling of the unit is not recommended as, for it to function satisfactorily, a large number of specialised tools are required, including a special hypoid housing spreader which has to be fitted before the differential unit can be removed from its housing.

8. DIFFERENTIAL INNER DRIVE SHAFTS - REMOVAL & REPLACEMENT

1. If oil is leaking from the inner drive shafts or if the ball bearings are worn or broken it will be necessary to remove the inner shafts and replace the defective components. Numerical references in brackets refer to Fig. 8.8.

2. Take off the main drive shaft, hub, and vertical links as described in Section 4. Thoroughly clean the area adjacent to the drive shaft and flanges

3. Drain the oil from the axle and with a $3/16$ in. Allen key undo the four Allen screws (16) which hold the seal housing plate (14) to the rear casing (11). It may not be possible to remove the screws from the housing at this stage because of masking by the drive flange, but all the screws can be undone completely. (See Fig. 8.8. on page 127).

125

CHAPTER EIGHT

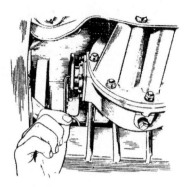

Fig. 8.6. Taking out the allen screws retaining the inner drive shaft.

4. The inner drive shaft can then be removed complete with the ball race (17), oil seal (15), and housing plate (14). (Fig. 8.8. refers).

Fig. 8.7. Withdrawing the inner drive shaft from hypoid casing.

5. Place the flange in the jaws of a vice and with a pair of circlip pliers take off the circlip (12) from the splined end of the shaft. Sometimes a washer will be found under the circlip. This must always be replaced.

6. The ball race is now carefully drifted off the axle shaft, followed by the housing plate complete with seal.

7. Carefully prise out the oil seal. Note that when fitting a new seal its lip must face inwards.

8. Replacement is a straightforward reversal of the removal sequence. Remember to fit the other screws loosely before fitting the housing to the drive shaft.

Fig. 8.8. EXPLODED VIEW OF THE REAR AXLE

1 Shims. 2 Differential side bearing. 3 Thrust washer. 4 Cross-shaft locking pin. 5 Differential sun gear. 6 Differential planet gear. 7 Thrust washer. 8 Paper gasket. 9 Rear mounting bolt. 10 Metalastik bush. 11 Hypoid rear casing. 12 Circlip. 13 Nyloc nut. 14 Seal housing plate. 15 Oil seal. 16 Allen screw. 17 Ball race. 18 Differential carrier. 19 Differential side bearing. 20 Shims. 21 Inner axle shaft and flange. 22 Nyloc nut. 23 Bolt. 24 Bolt. 25 Shim. 26 Rubber sealing ring. 27 Nylon bush. 28 Shim. 29 Stud. 30 Hub. 31 Nyloc nut. 32 Grease trap. 33 Outer seal housing. 34 Oil seal. 35 Ball race. 36 Paper gasket. 37 Trunnion housing. 38 Distance tube. 39 Grease plug. 40 Needle roller bearing. 41 Trunnion bearing inner oil seal. 42 Key. 43 Outer axle shaft. 44 Grease thrower disc. 45 Universal joint assembly. 46 Circlip. 47 Bearing cap. 48 Tubular dowel. 49 Bolt. 50 Mounting rubber. 51 Nyloc nut. 52 Plain washer. 53 Rubber pad. 54 Bolt. 55 Split pin. 56 Slotted nut. 57 Coupling flange. 58 Oil seal. 59 Pinion tail bearing. 60 Shims. 61 Spacer. 62 Mounting plate. 63 Bolt. 64 Hypoid nose piece casing. 65 Pinion lead bearing. 66 Spacer. 67 Pinion. 68 Crown wheel. 69 Differential cross shaft.

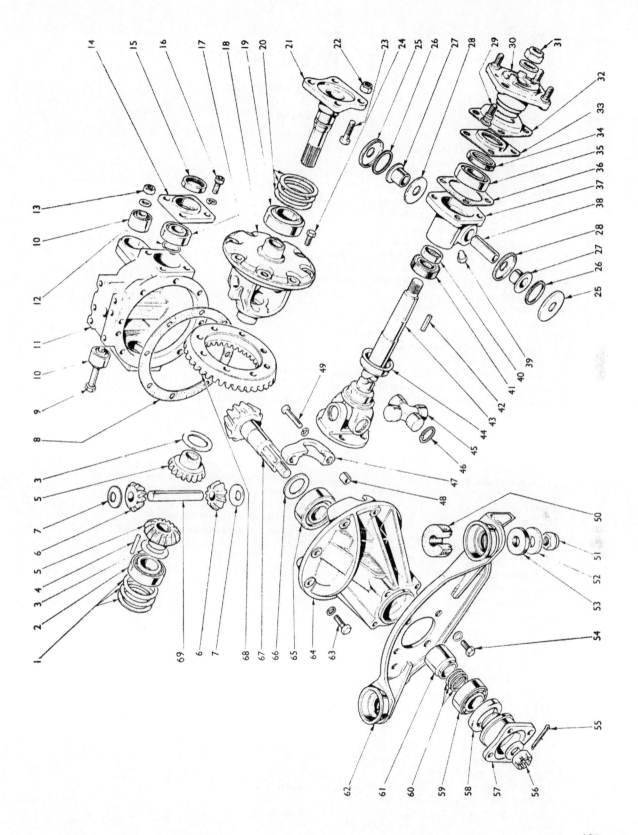

127

CHAPTER NINE

BRAKING SYSTEM

CONTENTS

Drum Brakes - General Description	1
Drum Brakes - Maintenance	2
Drum Brakes - Adjustment	3
Bleeding the Hydraulic System	4
Drum Brake Shoe - Inspection, Removal & Replacement	5
Flexible Hose - Inspection, Removal & Replacement	6
Brake Seals - Inspection & Overhaul	7
Front Wheel Cylinders - Removal & Replacement	8
Rear Wheel Cylinders - Removal & Replacement	9
Brake Master Cylinder - Removal & Replacement	10
Brake Master Cylinder - Dismantling & Reassembly	11
Handbrake Adjustment	12
Brake Pedal - Removal & Replacement	13
Front Brake Backplate - Removal & Replacement	14
Rear Brake Backplate - Removal & Replacement	15
Disc Brakes - General Description	16
Disc Brakes - Maintenance	17
Disc Brake Friction Pad - Inspection, Removal & Replacement	18
Brake Calliper - Removal, Dismantling & Reassembly	19
Front hub - removal, overhaul and refitting	20

SPECIFICATIONS

Make & Type
- 948 c.c. Heralds & Herald 1200 ... Girling drum brakes, front and rear
- Herald 1200 (optional) 12/50 & 13/60 ... Girling disc at front, drum at rear
- Footbrake ... Hydraulic on all 4 wheels
- Handbrake ... Mechanical to rear wheels only.
- Brake fluid ... Any make conforming to S.A.E. 70R3
- Front brakes - Drum diameter ... 8 in. (20.32 cm.)
-              - Disc diameter ... 9 in. (22.86 cm.)
- Shoe lining material (drum) ... Ferodo M.S.I.
- Disc pad material
  - (with 12P callipers) ... Don 55
  - (with 14LF callipers) ... Don 212

Rear Brakes - Drum Diameter ... 7 in.
- Lining material
- Cars with 12P callipers on front brakes and all 948 c.c. Heralds ... Ferodo M.S.I.
- Cars with 14 LF callipers on front brakes ... Don 242

Total Swept Area
- 948 c.c. Heralds & Herald 1200 with drum brakes ... 118 sq. in.
- Herald 1200 (discs) 12/50 & 13/60 with 12P callipers ... 199 sq. in.
- Herald 1200 (discs) 12/50 & 13/60 with 14 LF callipers ... 205 sq. in.

BRAKING SYSTEM

1. **DRUM BRAKES - GENERAL DESCRIPTION**

    The four wheel drum brakes fitted to 948 c.c. and certain 1200 models are of the internal expanding type and are operated hydraulically by means of the brake pedal which is coupled to the brake master cylinder and hydraulic fluid reservoir mounted on the front bulkhead.

    The front brakes are of the two leading shoe type with a separate cylinder for each shoe. Both cylinders are fixed to the backplate and the trailing end of each shoe is free to slide laterally in a small groove in the closed end of the brake cylinders, so ensuring automatic centralisation when the brakes are applied.

    The rear brakes are of the single leading shoe type, with one brake cylinder per wheel for both shoes. The cylinder is free to float on the backplate. Attached to each of the rear wheel operating cylinders is a mechanical expander operated by the handbrake lever through a cable which runs from the brake lever to a compensator on the rear axle and thence to the wheel operating levers.

    Drum brakes have to be adjusted periodically to compensate for wear in the linings. It is unusual to have to adjust the handbrake system as the efficiency of this system is largely dependent on the condition of the brake linings and the adjustment of the brake shoes. The handbrake can, however, be adjusted separately to the footbrake operated hydraulic system.

    The hydraulic brake system functions in the following manner:- On application of the brake pedal, hydraulic fluid under pressure is pushed from the master cylinder to the brake operating cylinders at each wheel, by means of a four way union and steel pipe lines and flexible hoses. (See Fig. 9.3.).

    The hydraulic fluid moves the pistons out so pushing the brake shoes into contact with the brake drums. This provides an equal degree of retardation on all four wheels in direct proportion to the brake pedal. Return springs between the backplate and the brake shoes draw the shoes together when the brake pedal is released.

2. **DRUM BRAKES - MAINTENANCE**

    1. Every 3,000 miles, carefully clean the top of the brake master cylinder reservoir, remove the cap, and inspect the level of the fluid which should be $\frac{1}{4}$ in. below the bottom of the filler neck. Check that the breathing holes in the cap are clear.
    2. If the fluid is below this level, top up the reservoir with Castrol Girling Amber Brake Fluid, or a fluid which conforms to specification SAE 70 R3. It is vital that no other type of brake fluid is used. Use of a non-standard fluid will result in brake failure caused by the perishing of the special seals in the master and brake cylinders. If topping up becomes frequent then check the metal piping and flexible hosing for leaks, and check for worn brake or master cylinders which will also cause loss of fluid.
    3. At intervals of 3,000 miles, or more frequently if pedal travel becomes excessive, adjust the brake shoes to compensate for wear of the brake linings.
    4. At the same time lubricate all joints in the handbrake mechanism with an oil can filled with Castrolite or similar.
    5. Every 6,000 miles apply grease to the handbrake cable guides and the compensator sector.

3. **DRUM BRAKES - ADJUSTMENT**
    1. Jack up one side of the car to attend to the brakes on that side.
    2. The brakes on all models are taken up by turning square headed adjusters on the rear of each backplate. The edges of the adjuster are easily burred if an ordinary spanner is used. Use a square headed brake adjusting spanner if possible. NOTE: When adjusting the rear brakes make sure the handbrake is off.

Fig. 9.1. The two square headed brake adjusters on each of the front brakes.

3. Two adjusters are fitted to each of the front wheels (see Fig. 9.1.) and one adjuster on the rear wheel backplate (see Fig. 9.2.).

Fig. 9.2. Turning the single brake adjuster on a rear wheel.

4. Turn the adjuster a quarter of a turn at a time until the wheel is locked. Then turn back the adjuster one notch so the wheel will rotate without binding.
5. Spin the wheel and apply the brakes hard to centralise the shoes. Recheck that it is not possible to turn the adjusting screw further without locking the shoe. NOTE A rubbing noise when the wheel is spun is usually due to dust in the brake drum. If there is no obvious slowing of the wheel due to brake binding there is no need to slacken off the adjusters until the noise disappears. Better to remove the drum and blow out the dust.

CHAPTER NINE

6. Repeat this process to the other three brake drums. A good tip is to paint the head of the adjusting screws white which will facilitate future adjustment by making the adjuster heads easier to see.

4. BLEEDING THE HYDRAULIC SYSTEM

1. Removal of all the air from the hydraulic system is essential to the correct working of the braking system, and before undertaking this examine the fluid reservoir cap to ensure that both vent holes, one on top and the second underneath but not in line, are clear; check the level of fluid and top up if required.
2. Check all brake line unions and connections for possible seepage, and at the same time check the condition of the rubber hoses, which may be perished.
3. If the condition of the wheel cylinders is in doubt, check for possible signs of fluid leakage.
4. If there is any possibility of incorrect fluid having been put into the system, drain all the fluid out and flush through with methylated spirits. Renew all piston seals and cups since there will be affected and could possibly fail under pressure.
5. Gather together a clean jam jar, a 9 in. length of tubing which fits tightly over the bleed nipples, and a tin of the correct brake fluid. (Girling Amber brake fluid).
6. To bleed the system clean the areas around the bleed valves, and start on the rear brakes first by removing the rubber cup over the bleed valve and fitting a rubber tube in position.
7. Place the end of the tube in a clean glass jar containing sufficient fluid to keep the end of the tube underneath during the operation.
8. Open the bleed valve with a spanner and quickly press down the brake pedal. After slowly releasing the pedal, pause for a moment to allow the fluid to recoup in the master cylinder and then depress again. This will force air from the system. Continue until no more air bubbles can be seen coming from the tube. At intervals make certain that the reservoir is kept topped up, otherwise air will enter at this point again.
9. Repeat this operation on all four brakes, and when completed, check the level of the fluid in the reservoir and then check the feel of the brake pedal, which should be firm and free from any 'spongy' action, which is normally associated with air in the system.

5. DRUM BRAKE SHOE - INSPECTION, REMOVAL & REPLACEMENT

After high mileages it will be necessary to fit replacement brake shoes with new linings. Refitting new brake linings to old shoes is not always satisfactory, but if the services of a local garage or workshop with brake lining equipment is available, then there is no reason why your own shoes should not be successfully relined.

1. Remove the hub cap, loosen off the wheel nuts, (photo), securely jack up the car, and remove the road wheel. Ensure the handbrake is off if the rear brake shoes are being removed.
2. Completely slacken off the brake adjustment and take out the two setscrews (photo) which hold the drum in place.
3. Remove the brake drum (photo). If it proves obstinate tap the rim gently with a soft headed hammer. The shoes are now exposed for inspection.
4. The brake linings should be renewed if they are so worn that the rivet heads are flush with the surface of the lining. If bonded linings are fitted they must be removed when the material has worn down to $\frac{1}{32}$ in. at its thinnest point. If the shoes are being removed to give access to the wheel cylinders, then cover the linings with masking tape (photo) to prevent any possibility of their becoming contaminated with grease or oil.
5. Press in each brake shoe steady pin securing washer against the pressure of its spring (photo).
6. Turn the head of the washer 90° so the slot will clear the securing bar on the steady pin and remove the spring and washer (photo).
7. Detach the shoes and return springs by pulling one end of the shoes away from the slot in the closed

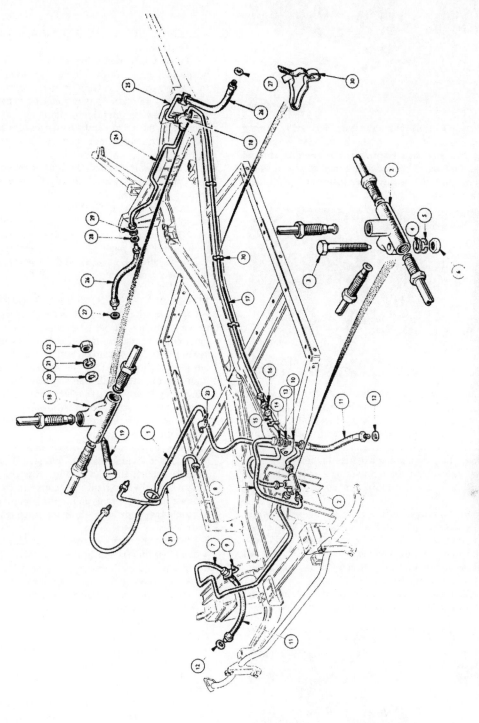

Fig. 9.3. EXPLODED VIEW OF THE HYDRAULIC PIPES, HOSES, AND CONNECTORS

1 Pipe from master cylinder to 4 way connector. 2 Four way connector. 3 Bolt. 4 Plain washer. 5 Lock washer. 6 Nut. 7 Pipe to R.H. front hose. 8 Pipe to L.H. front hose. 9 Hose mounting bracket R.H. 10 Hose mounting bracket L.H. 11 High pressure front hoses. 12 Gasket. 13 Shakeproof washer. 14 Nut. 15 Pipe assembly. 16 Connector. 17 Pipe. 18 Three way connector. 19 Bolt. 20 Plain washer. 21 Lock washer. 22 Nut. 23 Clip. 24 R.H. pipe. 25 L.H. rear brake pipe. 26 Rear brake hoses. 27 Gasket. 28 Shakeproof washer. 29 Nut. 30 Clip. 31 Hydraulic clutch pipe from master to slave cylinders

CHAPTER NINE

end of one of the brake cylinders (photo).

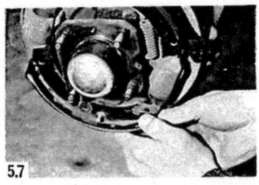
5.7

8. Disengage the brake shoe from the return spring (photo) carefully noting the holes into which the spring fits and then remove the remaining shoe in similar fashion. Place rubber bands over the wheel cylinders to prevent any possibility of the pistons dropping out.

5.8

9. Thoroughly clean all traces of dust from the shoes, backplates, and brake drums with a dry paint brush and compressed air, if available. Brake dust can cause squeal and judder and it is therefore important to clean out the brakes thoroughly.
10. Check that the pistons are free in their cylinders and that the rubber dust covers are undamaged and in position and that there are no hydraulic fluid leaks.
11. Prior to reassembly smear a trace of white brake grease to all sliding surfaces. The shoes should be quite free to slide on the closed end of the cylinder and the piston anchorage point. It is vital that no grease or oil comes in contact with the brake drums or the brake linings.
12. Replacement is a straightforward reversal of the removal procedure, but note the following points:-

5.12

a) Check that when the micram adjusters are replaced they are backed right off.
b) Do not omit to fit the steady pins and inner and outer washers (photo) if they were removed.
c) Ensure that the return springs are in their correct holes in the shoes and lie between them and the backplate.

6. FLEXIBLE HOSE - INSPECTION, REMOVAL & REPLACEMENT

Inspect the condition of the flexible hydraulic hoses leading from the chassis mounted metal pipes to the brake backplates. If any are swollen, damaged, cut, or chafed, they must be renewed.
1. Unscrew the metal pipe union nuts from its connection to the hose, and then holding the hexagon on the hose with a spanner, unscrew the attachment nut and washer.
2. The chassis end of the hose can now be pulled from the chassis mounting bracket and will be quite free.
3. Disconnect the flexible hydraulic hose at the backplate by unscrewing it from the brake cylinder. NOTE when releasing the hose from the backplate, the chassis end must always be freed first.
4. Replacement is a straightforward reversal of the above procedure.

7. BRAKE SEALS - INSPECTION & OVERHAUL

If hydraulic fluid is leaking from one of the brake cylinders it will be necessary to dismantle the cylinder and replace the dust cover and piston sealing rubber. If brake fluid is found running down the side of the wheel, or it is noticed that a pool of liquid forms alongside one wheel and the level in the master cylinder has dropped, and the hoses are all in good order, proceed as follows:-
1. Remove the brake drums and brake shoes as described in Section 5.
2. Ensure that all the other wheels, and all the other brake drums are in place. Remove piston sealing rubber and the spring from the leaking cylinder by applying gentle pressure to the footbrake. Place a quantity of rag under the backplate or a tray to catch the hydraulic fluid as it pours out of the cylinder.
3. Inspect the inside of the cylinder for score marks caused by impurities in the hydraulic fluid. If any are found the cylinder and piston will require renewal together as an exchange assembly.
4. If the cylinder is sound thoroughly clean it out with fresh hydraulic fluid.
5. The old rubber seal will probably be swollen and visibly worn. Smear the new rubber seal with hydraulic fluid and reassemble in the cylinder the spring, seal and piston, and then the rubber boot. The seal must be fitted with its lip towards the bottom of the cylinder.
6. Replenish the brake fluid, replace the brake shoes and brake drum, and bleed the hydraulic system as previously described.

8. FRONT WHEEL CYLINDERS - REMOVAL & REPLACEMENT

1. Remove the appropriate front brake drum and brake shoes as described in Section 5.

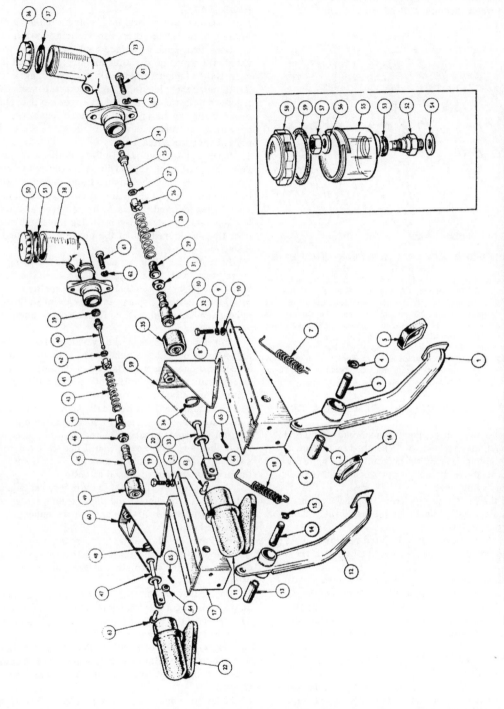

Fig. 9.4. EXPLODED VIEW OF THE BRAKE AND CLUTCH MASTER CYLINDERS, PEDALS AND BRACKETS

1 Clutch pedal. 2 Bush. 3 Shaft. 4 Circlip. 5 Rubber pad. 6 Bracket. 7 Spring. 8 Bolt. 9 Lock washer. 10 Plain washer. 11 Dust cover. 12 Pedal. 13 Bush. 14 Shaft. 15 Circlip. 16 Rubber pad. 17 Bracket. 18 Return spring. 19 Bolt. 20 Lock washer. 21 Plain washer. 22 Dust cover. 23 Clutch master cylinder body. 24 Valve seal. 25 Valve stem. 26 Valve spacer. 27 Spring washer. 28 Plunger spring. 29 Spring retainer. 30 Cylinder plunger. 31 Seal. 32 Taper seal. 33 Push rod. 34 Circlip. 35 Dust cover. 36 Cap. 37 Gasket. 38 Master cylinder body. 39 Seal. 40 Valve stem. 41 Valve spacer. 42 Spring washer. 43 Spring. 44 Retainer. 45 Plunger. 46 Gland seal. 47 Pushrod. 48 Circlip. 49 Dust cover. 50 Cap. 51 Gasket. 52 Adaptor. 53 Gasket. 54 Seal. 55 Reservoir. 56 Washer plain. 57 Nut. 58 Filler cap. 59 Support bracket. 60 Brake master cylinder support bracket. 61 Bolt. 62 Lock washer. 63 Pin. 64 Plain washer. 65 Cotter pin.

CHAPTER NINE

2. Undo the bridge pipe unions from the cylinders and also the flexible pipe if this is fitted to the cylinder being removed.
3. Undo the two nuts and washers which hold each wheel cylinder in place and remove the cylinder.
4. Replacement is a straightforward reversal of the dismantling process. Tighten the wheel cylinder nuts, and when assembled bleed the brakes.

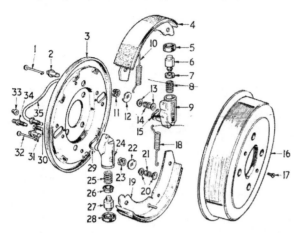

Fig. 9.5. EXPLODED VIEW OF LEFT HAND FRONT DRUM BRAKE ASSEMBLY

1 Brake shoe steady pin. 2 Adjuster shank. 3 Brake backplate. 4 Brake shoe. 5 Dust cover. 6 Piston. 7 Seal. 8 Spring. 9 Wheel cylinder assembly. 10 Brake shoe return spring. 11 Spring. 12 Adjuster cam. 13 Steady pin cups. 14 Spring. 15 Seal. 16 Brake drum. 17 Countersunk screw. 18 Brake shoe return spring. 19 Brake shoe. 20 Steady pin cups. 21 Spring. 22 Adjuster cam. 23 Spring. 24 Wheel cylinder. 25 Spring. 26 Seal. 27 Piston. 28 Dust cover. 29 Seal. 30 Screw. 31 Adjuster shank. 32 Brake shoe steady pin. 33 Dust cap. 34 Bleed screw. 35 Bridge pipe.

9. REAR WHEEL CYLINDERS - REMOVAL & REPLACEMENT

1. Remove the left or right-hand brake drum and brake shoes as required, as described in Section 5. To avoid having to drain the hydraulic system screw down the master cylinder reservoir cap tightly over a piece of cellophane.
2. Free the hydraulic pipe from the wheel cylinder at the union, and disconnect the handbrake cable clevis from its lever.
3. Take off the dust excluder, the retaining plate and the spring clip, and remove the cylinder from the backplate.
4. On replacement smear the slot in the backplate and the cylinder neck with Girling white brake grease. The rest of the replacement process is a straightforward reversal of the removal sequence. Bleed the brakes on completion of reassembly.

10. BRAKE MASTER CYLINDER - REMOVAL & REPLACEMENT

1. To remove the Girling type C.V. brake master cylinder first undo the union from the brake pipe outlet and pull the pipe clear.
2. Pull back the rubber dust excluder and free the clevis pin from the yoke on the end of the pushrod and then undo the two bolts and spring washers from the bracket on which the master cylinder flange is mounted.
3. The master cylinder can now be pulled off its bracket and dismantled further if required. Replacement is a straightforward reversal of the removal sequence.

11. BRAKE MASTER CYLINDER - DISMANTLING & REASSEMBLY

1. First detach the fluid line, using a blanking plug in the pipe to prevent dirt from entering the system.
2. Note that when a replacement cylinder is to be fitted, the working surfaces are protected and it is essential to lubricate the seals before fitting.
3. Remove the blanking plugs from the pipe line, together with the pushrod dust cover so that clean brake fluid can be injected at these locations. By operating the piston several times the fluid will spread over the surfaces.
4. If the master cylinder is to be dismantled after removal, pull back the pushrod cover and remove the circlip so that the pushrod and dished washer can be pulled out. This will expose the plunger with a seal attached, and this must be removed as a unit. The assembly is separated by lifting the thimble leaf over the shouldered end of the plunger. The seal is then eased off.
5. Depress the plunger return spring allowing the valve stem to slide through the keyhole in the thimble, thus releasing the tension in the spring.
6. Detach the valve spacer taking care of the spacer spring washer which will be found located under the valve head.
7. Examine the bore of the cylinder carefully for any scores or ridges, and if this is found to be smooth all over, new seals can be fitted. If there is any doubt of the condition of the bore then a new cylinder must be fitted.
8. If examination of the seals shows them to be apparently oversize, or very loose on the plunger, suspect oil contamination in the system. Oil will swell these rubber seals, and if one is found to be swollen, it is reasonable to assume that all seals in the braking system will need attention.
9. To reassemble the master cylinder, replace the old valve seal as shown, and then replace the spring washer with its domed side against the underside of the valve head.
10. Replace the plunger return spring centrally on the spacer, insert the thimble into the spring and depress until the valve stem engages in the keyhole of the thimble.
11. Check that the spring is central on the spacer before refitting a new plunger seal onto the plunger with the flat face against the face of the plunger, and a new back seal if required.
12. Insert the reduced end of the plunger into the thimble until the thimble engages under the shoulder of the plunger and press home the thimble leaf as shown.
13. Make sure that the bore is clean, smear the plunger with brake fluid and insert the assembly into the bore valve end first, easing the lips of the plunger seal carefully into the bore.
14. Replace the pushrod and refit the circlip into the groove in the cylinder body, and replace the rubber cover.

BRAKING SYSTEM

12. HANDBRAKE ADJUSTMENT

1. If the handbrake is in need of adjustment, excessive travel of the lever is taken up automatically when the rear brakes are adjusted. After high mileages it is possible that the handbrake cables will have stretched and will need to be adjusted. Numerical references in brackets refer to Fig. 9.7.

2. With the rear wheels free off the ground and chocks under the front wheels to prevent any forward movement, ensure the handbrake is fully off, and then lock each brake drum by screwing the brake adjusters in as far as possible.

3. Disconnect the pull off spring (12) from the rear of the backplate, remove the split pin from the clevis pin (17) and pull out the clevis pin to free the clevis (16) from the brake lever.

4. Loosen the locknut (18) and screw the clevis onto the threaded rod until the clevis pin can be fitted to the clevis and the brake lever without pulling the lever from the off position or straining the cable.

5. Resecure the clevis, reconnect the spring and then adjust the cable bracket (20) so there is slight tension in the spring (12).

6. Slacken off the rear brake adjusters in the normal way until adjustment is correct. Lower the car to the ground.

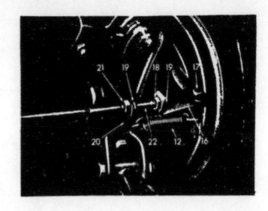

Fig. 9.6. View of the handbrake adjustment mechanism.

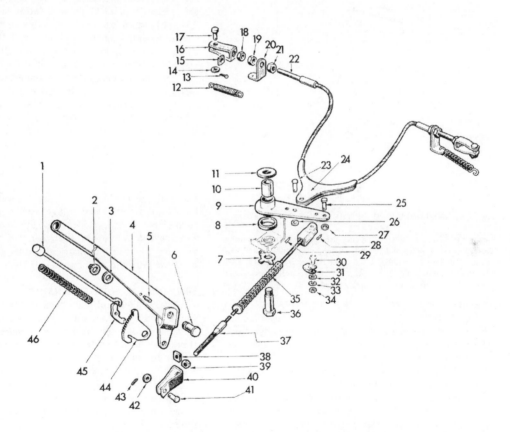

Fig. 9.7. EXPLODED VIEW OF THE HANDBRAKE MECHANISM

1 Pawl release rod. 2 Circlip. 3 Plain washer. 4 Handbrake lever. 5 Pawl pivot pin. 6 Pivot pin. 7 Lock plate. 8 Rubber seal. 9 Relay lever. 10 Bush. 11 Felt seal. 12 Pull-off spring. 13 Split pin. 14 Plain washer. 15 Square nut. 16 Clevis. 17 Clevis pin. 18 Locknut. 19 Adjusting nut. 20 Adjustable spring anchor. 21 Locknut. 22 Secondary cable. 23 Clevis pin. 24 Compensator sector. 25 Clevis pin. 26 Plain washer. 27 Plain washer. 28 Split pin. 29 Split pin. 30 Clamp bolt. 31 Clamp. 32 Plain washer. 33 Spring washer. 34 Nut. 35 Spring. 36 Pivot bolt. 37 Primary cable. 38 Square nut. 39 Locknut. 40 Clevis. 41 Clevis pin. 42 Plain washer. 43 Split pin. 44 Ratchet. 45 Pawl. 46 Pawl spring.

135

CHAPTER NINE

13. BRAKE PEDAL - REMOVAL & REPLACEMENT

1. If it is wished to renew the pivot bush or to remove the pedal first pull back the rubber dust excluder and then pull out the clevis pin.
2. Pull off the pedal return spring and with a pair of circlip pliers remove the circlip from the end of the pivot pin.
3. Push out the pivot pin from the bracket and pedal and take the pedal out of the bracket.
4. The old pivot bush if worn is simply pushed out of the pedal and a new one slid into place. Reassembly is a straightforward reversal of the removal sequence.

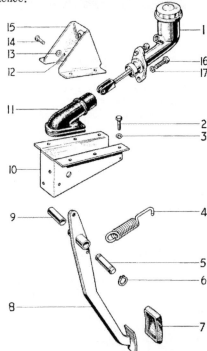

Fig. 9.8. EXPLODED VIEW OF THE BRAKE PEDAL AND BRACKET

1 Master cylinder. 2 Bolt. 3 Spring washer. 4 Return spring. 5 Pivot pin. 6 Circlip. 7 Pedal rubber. 8 Pedal. 9 Pedal pivot bush. 10 Pedal bracket. 11 Rubber dust excluder. 12 Split pin. 13 Plain washer. 14 Clevis pin. 15 Master cylinder bracket. 16 Bolt. 17 Spring washer.

14. FRONT BRAKE BACKPLATE - REMOVAL & REPLACEMENT

1. Loosen the road wheel securing nuts, jack up the front of the car, remove the road wheel and then the brake drum.
2. Disconnect the flexible hydraulic hose from the backplate as described in Section 6.
3. Prise off the hub cap, pull out the split pin, and undo and remove the castellated nut and washer and pull the hub complete with bearings off the stub axle.
4. Knock back the locking tabs from the four bolts which hold the backplate in place against the vertical link and steering arm, undo the bolts and pull the brake backplate off the stub axle.
5. Replacement is a straightforward reversal of the removal sequence. Always fit new tab washers.

15. REAR BRAKE BACKPLATE - REMOVAL & REPLACEMENT

1. Loosen the road wheel securing nuts, jack up the rear of the car, remove the road wheel and then the brake drum.
2. Disconnect the flexible hydraulic hose from the backplate as described in Section 6.
3. Pull off the rear hub as described in Chapter 8, Section 5.
4. Disconnect the handbrake cable clevis from the lever on the backplate. Knock back the tabs which hold the securing bolts in place, and undo the latter. The backplate can now be pulled away from the trunnion housing.
5. Replacement is a straightforward reversal of the removal sequence.

16. DISC BRAKES - GENERAL DESCRIPTION

Disc brakes are fitted to the front wheels of later models of the 1200 and all models of the 13/60 which retain the previous system of single leading shoe drum brakes for the rear wheels, together with the mechanically operated handbrake.

The brakes fitted to the front two wheels are of the rotating disc and static calliper type, with one calliper per disc, each calliper containing two piston operated friction pads, which on application of the footbrake pinch the disc rotating between them.

Application of the footbrake creates hydraulic pressure in the master cylinder and fluid from the cylinder travels via steel and flexible pipes to the cylinders in each half of the callipers, the fluid so pushing the pistons, to which are attached the friction pads, into contact with either side of each disc.

Two rubber seals are fitted to the operating cylinders. The outer seal prevents moisture and dirt from entering the cylinder, while the inner seal, which is retained in a groove inside the cylinder, prevents fluid leakage, and provides a running clearance for the pad irrespective of how worn it is, by moving it back a fraction when the brake pedal is released.

As the friction pad wears so the pistons move further out of the cylinders and the level of the fluid in the hydraulic reservoir drops, but disc pad wear is thus taken up automatically and eliminates the need for periodic adjustments by the owner.

17. DISC BRAKES - MAINTENANCE

Every 3,000 miles, check the level of the fluid in the master cylinder reservoir as detailed in the section headed 'Drum Brakes - Maintenance'. In this case, however, use Girling disc brake fluid, or if this is not available, a fluid to the specification S.A.E. 70 R3, for replenishment purposes.

At the same time examine the wear in the brake disc pads and change them round if one is very much more worn on one side of the rotating disc than on the other.

18. DISC BRAKE FRICTION PAD - INSPECTION, REMOVAL & REPLACEMENT

1. Remove the front wheels and inspect the amount of friction material left on the friction pads. The pads must be renewed when the thickness of the material has worn down to $\frac{1}{16}$ in.
2. Referring to Fig. 9.9. pull out the wire clips (8) which secure the pad retaining pins (7) in place, and remove the pins.

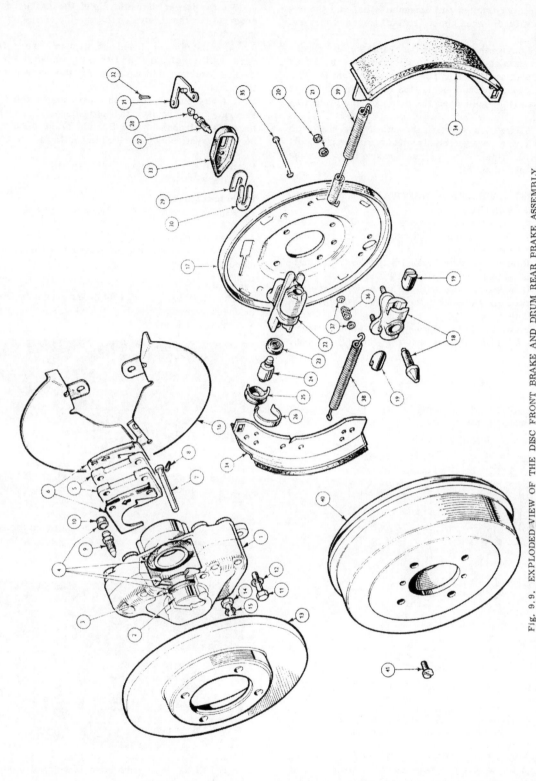

Fig. 9.9. EXPLODED VIEW OF THE DISC FRONT BRAKE AND DRUM REAR BRAKE ASSEMBLY
1 Front brake calliper. 2 Piston. 3 Sealing ring. 4 Service kit. 5 Lining. 6 Anti-squeal shim. 7 Pin. 8 Clip. 9 Bleed screw. 10 Cover. 11 Bolt. 12 Spring washer. 13 Friction disc. 14 Bolt. 15 Spring washer. 16 Dust shield. 17 Backplate. 18 Adjuster unit. 19 Tappet. 20 Nut. 21 Spring washer. 22 Wheel cylinder. 23 Seal. 24 Piston. 25 Dust cover. 26 Retainer. 27 Bleed screw. 28 Dust cap. 29 Plate retaining wheel cylinder. 30 Spring plate. 31 Operating lever assembly. 32 Cotter pin. 33 Dust cover. 34 Brake shoe assembly. 35 Shoe hold down pin. 36 Spring. 37 Cup washer. 38 Return spring. 39 Return spring (cylinder end). 40 Rear brake drum. 41 Screw.

137

CHAPTER NINE

3. The friction pads (5) and anti-squeal shims (6) can now be lifted from the calliper.

4. Carefully clean the recesses in the calliper in which the friction pad assemblies lie, and the exposed face of each piston from all traces of dirt and rust.

5. Remove the cap from the hydraulic fluid reservoir and place a large rag underneath the unit. Press the pistons in each half of the calliper right in - this will cause the fluid level in the reservoir to rise and possibly to spill over the brim onto the protective rag.

6. Fit new pads and refit the anti-squeal shims with the arrow towards the direction of rotation. Insert the pad retainer pins (7) and secure them with the retainer clips (8).

19. BRAKE CALLIPER - REMOVAL, DISMANTLING & REASSEMBLY

1. Jack up the car, remove the road wheel, and disconnect the flexible hydraulic pipe as previously detailed in Section 6.

2. Remove the disc brake friction pads and anti-squeal shims as previously described.

3. Unscrew the two calliper mounting bolts and lockwashers and remove the calliper assembly from the disc.

4. Referring to Fig. 9.10, pull off the dust covers (8) and remove the pistons from the calliper body (3).

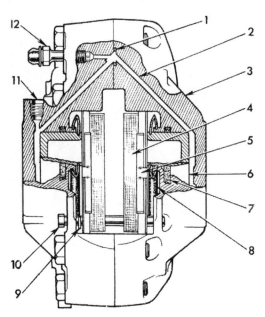

Fig. 9.10. A SECTIONED VIEW OF THE CALLIPER ASSEMBLY
1 Rubber 'O' ring. 2 Fluid transfer channels. 3 Calliper body. 4 Brake pad. 5 Anti-squeal plate. 6 Piston. 7 Piston sealing ring. 8 Dust cover. 9 Retaining clip. 10 Retaining pin. 11 Flexible hose connection. 12 Bleed nipple.

5. Very carefully remove the rubber piston sealing rings (7) from their recesses in the calliper. The pistons, cylinders, and rubbers should be cleaned only with clean brake fluid.

6. Examine the components carefully, renew the rubbers as a matter of course, and replace the pistons if slightly grooved or otherwise worn.

7. Reassembly commences by carefully fitting new piston sealing rings into the recesses in the calliper cylinders.

8. Fit the larger diameter lip of the rubber dust cover (8) to the groove on the outside of the top of the cylinder.

9. Slide the pistons closed end first into the cylinders, with great care, and then fit the outer lip of the dust excluder into the groove in the outer end of the piston.

10. Fit the calliper over the disc, insert the two securing bolts, replace the anti-squeal shims and the pads. Reconnect the flexible brake hose and bleed the system as described in Section 4.

20. FRONT HUB - REMOVAL, OVERHAUL AND REFITTING

Drum brake models
1. Remove the brake drum as described in Section 5.

Disc brake models
2. Remove the calliper unit as described in Section 19, but do not dismantle it.

All models
3. The grease cap is a very tight fit and is removed from the hub by screwing in the No. 10 UNF setscrew supplied with the vehicle tool kit.

20.4

4. Pull out the split pin from the castellated nut (photo), and undo and remove the castle nut, together with the washer.

20.5

5. The hub can then be pulled from the stub axle, together with its bearings (photo).

Disc brake models
6. Before separating the disc and hub, scribe an alignment mark to facilitate reassembly.

138

BRAKING SYSTEM

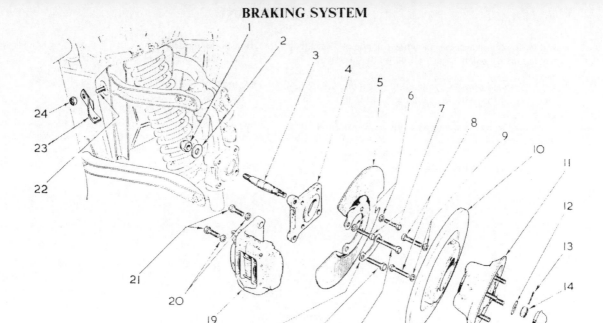

Fig. 9.11. EXPLODED VIEW OF THE FRICTION DISC AND HUB COMPONENTS
1 Nyloc nut. 2 Washer. 3 Stub axle. 4 Calliper mounting bracket. 5 Disc shield. 6 Spring washers. 7 Bolts. 8 Bolts. 9 Spring washers. 10 Disc. 11 Hub. 12 Washer. 13 Split pin. 14 Slotted nut. 15 Cap. 16 Bolt. 17 Bolt. 18 Lockplate. 19 Calliper assembly. 20 Spring washers. 21 Bolts. 22 Bolt. 23 Bracket. 24 Nyloc nut.

7. Separate the disc and hub by removing the bolts and spring washers and pulling the two items apart.

9. Clean the hub and ensure no burrs are present.
10. Commence reassembly by driving in the new bearing outer races so that the tapers are facing outwards (photo).

Disc brake models
11. Ensure the mating faces of the hub and disc are clean and reassemble, making sure the scribe marks are aligned.

All models
12. Fit the inner races to the hub (photo) and slide the hub assembly onto the stub axle as far as it will go.
13. Refit the washer and finger-tighten the castellated nut. Turn back the castellated nut to align with the first split pin hole in the stub axle and then measure the hub endfloat. This should be between 0.003 and 0.005 in (0.076 and 0.127 mm). If the endfloat is correct, centre punch alignment marks on the end of the stub axle and castellated nut. If the endfloat is incorrect, remove the castellated nut and file the rear face until the correct endfloat is obtained.
14. Remove the hub from the stub axle and lubricate the bearings with a lithium-based grease, and fit a new hub oil

All models
8. Only fit new bearings as complete sets. The old bearing outer races, oil seal and retainer can be driven out of the hub using a metal rod (photo).

139

CHAPTER NINE

seal to the seal retainer using jointing compound. When the seal has dried, soak in engine oil and gently squeeze out the surplus.

15. With the bearings in position, fit the seal retainer to the inner end of the hub with the oil seal facing into the hub centre.

16. Slide the hub assembly onto the stub axle, refit the washer and castellated nut and tighten the nut until the punch marks are in alignment. Fit a new split pin to lock the castellated nut and refit the grease cap.

Drum brake models
17. Refit the brake drum as described in Section 5.

Disc brake models
18. Refit the calliper as described in Section 19.
19. Measure the run-out of the disc at its outer periphery using feeler gauges positioned between the calliper and disc. If run-out exceeds 0.002 in (0.0508 mm), remove the disc and re-position it on the hub. If run-out still exceeds the limit stated, the disc is probably distorted and will require renewal.

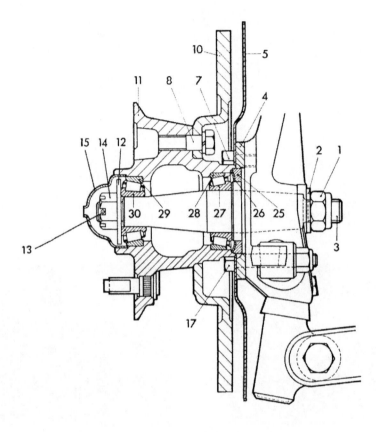

FIG. 9.12. SECTION THROUGH HUB
1 Nyloc nut. 2 Washer. 3 Stub axle. 4 Calliper mounting bracket. 5 Disc shield. 7 Bolts. 8 Bolts. 10 Disc. 11 Hub. 12 Washer. 13 Split pin. 14 Slotted nut. 15 Cap. 25 Hub oil seal. 26 Seal retainer. 27 Bearing inner race. 28 Bearing outer race. 29 Bearing outer race. 30 Bearing inner race.

BRAKING SYSTEM
FAULT FINDING CHART

Cause	Trouble	Remedy
SYMPTOM:	**PEDAL TRAVELS ALMOST TO FLOORBOARDS BEFORE BRAKES OPERATE**	
Leaks and air bubbles in hydraulic system	Brake fluid level too low	Top up master cylinder reservoir. Check for leaks.
	Wheel cylinder leaking	Dismantle wheel cylinder, clean, fit new rubbers and bleed brakes.
	Master cylinder leaking (Bubbles in master cylinder fluid)	Dismantle master cylinder, clean, and fit new rubbers. Bleed brakes.
	Brake flexible hose leaking	Examine and fit new hose if old hose leaking. Bleed brakes.
	Brake line fractured	Replace with new brake pipe. Bleed brakes.
	Brake system unions loose	Check all unions in brake system and tighten as necessary. Bleed brakes.
Normal wear	Linings over 75% worn	Fit replacement shoes and brake linings.
Incorrect adjustment	Brakes badly out of adjustment	Jack up car and adjust brakes.
	Master cylinder push rod out or adjustment causing too much pedal free movement	Reset to manufacturer's specification.
SYMPTOM:	**BRAKE PEDAL FEELS SPRINGY**	
Brake lining renewal	New linings not yet bedded-in	Use brakes gently until springy pedal feeling leaves.
Excessive wear or damage	Brake drums badly worn and weak or cracked	Fit new brake drums.
Lack of maintenance	Master cylinder securing nuts loose	Tighten master cylinder securing nuts. Ensure spring washers are fitted.
SYMPTOM:	**BRAKE PEDAL FEELS SPONGY & SOGGY**	
Leaks or bubbles in hydraulic system	Wheel cylinder leaking	Dismantle wheel cylinder, clean, fit new rubbers, and bleed brakes.
	Master cylinder leaking (Bubbles in master cylinder reservoir)	Dismantle master cylinder, clean, and fit new rubbers and bleed brakes. Replace cylinder if internal walls scored.
	Brake pipe line or flexible hose leaking	Fit new pipeline or hose.
	Unions in brake system loose	Examine for leaks, tighten as necessary.
SYMPTOM:	**EXCESSIVE EFFORT REQUIRED TO BRAKE CAR**	
Lining type or condition	Linings badly worn	Fit replacement brake shoes and linings.
	New linings recently fitted - not yet bedded-in	Use brakes gently until braking effort normal.
	Harder linings fitted than standard causing increase in pedal pressure	Remove linings and replace with normal units.
Oil or grease leaks	Linings and brake drums contaminated with oil, grease, or hydraulic fluid	Rectify source of leak, clean brake drums, fit new linings.
SYMPTOM:	**BRAKES UNEVEN & PULLING TO ONE SIDE**	
Oil or grease leaks	Linings and brake drums contaminated with oil, grease, or hydraulic fluid	Ascertain and rectify source of leak, clean brake drums, fit new linings.
Lack of maintenance	Tyre pressures unequal	Check and inflate as necessary.
	Radial ply tyres fitted at one end of car only	Fit radial ply tyres of the same make to all four wheels.
	Brake backplate loose	Tighten backplate securing nuts and bolts.
	Brake shoes fitted incorrectly	Remove and fit shoes correct way round.
	Different type of linings fitted at each wheel	Fit the linings specified by the manufacturers all round.
	Anchorages for front suspension or rear axle loose	Tighten front and rear suspension pick-up points including spring anchorage.
	Brake drums badly worn, cracked or distorted	Fit new brake drums.

BRAKING SYSTEM

Cause	Trouble	Remedy
SYMPTOM:	BRAKES TEND TO BIND, DRAG, OR LOCK-ON	
Incorrect adjustment	Brake shoes adjusted too tightly Handbrake cable over-tightened Master cylinder push rod out of adjustment giving too little brake pedal free movement	Slacken off brake shoe adjusters two clicks. Slacken off handbrake cable adjustment. Reset to manufacturer's specifications.
Wear or dirt in hydraulic system or incorrect fluid	Reservoir vent hole in cap blocked with dirt Master cylinder by-pass port restricted – brakes seize in 'on' position Wheel cylinder seizes in 'on' position	Clean and blow through hole. Dismantle, clean, and overhaul master cylinder. Bleed brakes. Dismantle, clean, and overhaul wheel cylinder. Bleed brakes.
Mechanical wear	Brake shoe pull off springs broken, stretched or loose	Examine springs and replace if worn or loose.
Incorrect brake assembly	Brake shoe pull off springs fitted wrong way round, omitted, or wrong type used	Examine, and rectify as appropriate.
Neglect	Handbrake system rusted or seized in the 'on' position	Apply 'Plus Gas' to free, clean and lubricate.

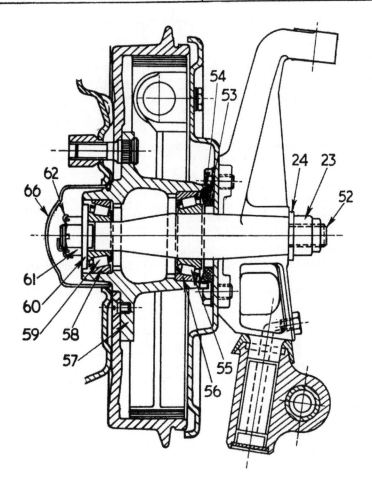

Fig. 9.13. CROSS SECTION OF HERALD 1200 DRUM BRAKE AND HUB ASSEMBLY
23 Nyloc nut. 24 Plain washer. 52 Stub axle. 53 Felt seal. 54 Seal retainer. 55 Taper roller bearing—inner. 56 Roller bearing —outer ring. 57 Hub. 58 Roller bearing—outer ring. 59 Taper roller bearing—outer. 60 'D' washer. 61 Slotted nut. 62 Split pin. 66 Cap.

CHAPTER TEN
ELECTRICAL SYSTEM

CONTENTS

General Description	1
Battery - Removal & Replacement	2
Battery - Maintenance & Inspection	3
Electrolyte Replenishment	4
Battery Charging	5
Dynamo - Routine Maintenance	6
Dynamo - Testing in Position	7
Dynamo - Removal & Replacement	8
Dynamo - Dismantling & Inspection	9
Dynamo - Repair & Reassembly	10
Starter Motor - General Description	11
Starter Motor - Testing in Engine	12
Starter Motor - Removal & Replacement	13
Starter Motor - Dismantling & Reassembly	14
Starter Motor Drive - General Description	15
Starter Motor Drive - Removal & Replacement	16
Starter Motor Bushes - Inspection, Removal & Replacement	17
Control Box - General Description	18
Cut-out & Regulator Contacts - Maintenance	19
Voltage Regulator Adjustment	20
Cut-out Adjustment	21
Flasher Circuit - Fault Tracing & Rectification	22
Windscreen Wiper Mechanism - Maintenance	23
Windscreen Wiper Arms - Removal & Replacement	24
Windscreen Wiper Mechanism - Fault Diagnosis & Rectification	25
Wiper Motor - Removal & Replacement	26
Wheelbox - Removal & Replacement	27
Windscreen Wiper Motor - Dismantling, Inspection & Reassembly	28
Horns - Fault Tracing & Rectification	29
Fuel Contents Gauge - Fault Tracing & Rectification	30
Fuel Gauge Sender Unit - Removal & Replacement	31
Temperature Gauge - Fault Tracing & Rectification	32
Headlamp Bulbs - Removal & Replacement - Focusing	33
Front, Side & Flasher Bulbs - Removal & Replacement	34
Stop, Rear & Flasher Bulbs - Removal & Replacement	35
Instrument Panel & Warning Bulbs - Removal & Replacement	36
Number Plate Bulb - Removal & Replacement	37
Fuse - Herald 13/60	38

SPECIFICATIONS

Battery	Lead acid.
Type	Lucas BT.7A (home).
	Lucas BTZ.7A (export).
Earthed terminal	Positive '+' (later 1200, 12/50 & 13/60 negative).
Capacity at 10 hr. rate	38 ampere hours.
Capacity at 20 hr. rate	43 ampere hours.
No. of plates per cell	7.
Electrolyte to fill one cell	1 pint (1.2 U.S. pints, .57 litres).

CHAPTER TEN

Specific gravity charged
    Climates below 32°C ... ... 1.270 to 1.290.
    Climates above 32°C ... ... 1.130 to 1.150.
Recharging current ... ... 5.0 amps.

Dynamo ... ... Lucas C40/1 (early models C39PV-2).
  Maximum output
    C39PV-2 ... ... 19 amps.
    C40/1 ... ... 22 amps.
  No. of brushes ... ... 2.
  Minimum brush length ... ... 11/32 in. (9 mm.).
  Brush Spring tension ... ... 22 to 25 ozs. (.62 to .71 kgs).
  Field resistance
    C39PV-2 ... ... 6.1 ohms.
    C40/1 ... ... 6.0 ohms.
  Cutting in speed
    C39PV-2 ... ... 1,050 to 1,200 r.p.m.
    C40/1 ... ... 1,250 to 1,450 r.p.m.

Starter Motor ... ... Lucas M.35 G.
  No. of brushes ... ... 4.
  Lock torque ... ... 7.2 to 7.6 lb/ft. at 320 to 340 amps.
  Torque at 1,000 r.p.m. ... ... 4.4 to 4.6 lb/ft. at 230 to 250 amps.
  Brush spring tension ... ... 32 to 40 ozs. (.9 to 1.1 kgs).

Regulator/Control Box ... ... Lucas RB.106/2.
  Cut in voltage ... ... 12.7 to 13.3 volts.
  Drop off voltage ... ... 8.5 to 11.0 volts.
  Open circuit voltage settings ... ... 10°C (50°F) 16.1 to 16.7 volts.
                                          20°C (68°F) 16.0 to 16.6 volts.
                                          30°C (86°F) 15.9 to 16.5 volts.
                                          40°C (104°F) 15.8 to 16.4 volts.
  Reverse current ... ... 3.5 to 5.0 amps.
  Current regulator ... ... 22 + or - 1 amp.

Fuses ... ... Adjacent to coil on bulkhead - 13/60 models only
  No. of fuses ... ... 1 - 25 amps.
  No. of fuses - all other models ... ... None.

Windscreen wiper motor ... ... Lucas DR.3A (early models Lucas DR.2).
  Normal running current ... ... 2.7 to 3.4 amps.
  Drive to wheelboxes ... ... Rack and cable.
  Armature endfloat ... ... .008 to .012 in. (.20 to .30 mm.).
  Armature resistance ... ... .29 to .352 ohms.
  Field resistance ... ... 8.0 to 9.5 ohms.
  Wiping speed ... ... 44 to 48 cycles per minute.

Horns
  948 c.c. Heralds ... ... Lucas 725H and 725L.
  Herald 1200, 12/50 and 13/60 ... ... Lucas 9H.
  Maximum current 725H and 725L ... ... 5 amps per horn.
             9H ... ... 3 1/2 amps per horn.

Bulbs	Volts	Watts	Part No.
Headlamps - Left-hand dip	12	50/40	508348
Right-hand dip	12	36/36	59469
Continental (Duplo)	12	45/50	501475
Vertical dip	12	35/35	60796
Side lamps	12	6	59467
Flashers	12	21	502379
Stop/Tail	12	21/6	502387
Plate illumination	12	6	501436
Panel and warning lamps	12	2.2	59492
Interior illumination - Amber	12	6	508997
Estate car	12	6	59897

ELECTRICAL SYSTEM

1 GENERAL DESCRIPTION

The electrical system fitted to all Heralds is of the 12-volt type and the major components comprise: A twelve-volt battery with positive earth on early models, later post 1967 1200 and 13/60 models have the negative terminal earthed; a control box and cut-out; a Lucas C39PU-2 or C40/1 dynamo which is fitted to the front left hand side of the engine and is driven by the fan belt from the crankshaft pulley wheel (these dynamos are identical for all practical purposes); and a starter motor which is fitted to the gearbox bellhousing on the left hand side of the engine. An unusual feature is that no fuses are fitted. The ignition system is also part of the electrical system but, because of its importance and complexity, is covered separately in Chapter 4.

On negative earth cars great care should be taken if fitting any electrical accessory containing silicon diodes or transistors that their polarity is correct. Serious damage may otherwise result to the components concerned.

Items such as radios, tape recorders, electronic ignition systems, electronic tachometers, automatic dipping, parking lamp and anti-dazzle mirrors should all be checked for correct polarity.

The seven plate per cell twelve-volt battery supplies a steady supply of current for the ignition, lighting, and other electrical circuits, and provides a reserve of electricity when the current consumed by the electrical equipment exceeds that being produced by the dynamo.

The dynamo is of the two brush type and works in conjunction with the voltage regulator and cut-out. The dynamo is cooled by a multi-bladed fan mounted behind the dynamo pulley, and blows air through cooling holes in the dynamo end brackets. The output from the dynamo is controlled by the voltage regulator which ensures a high output if the battery is in a low state of charge or the demands from the electrical equipment high, and a low output if the battery is fully charged and there is little demand from the electrical equipment.

2. BATTERY - REMOVAL & REPLACEMENT

1. The earthed battery terminal lead should always be removed first. Therefore on later negative earth cars remove the negative lead before the positive, and replace the negative lead last. On positive earth models disconnect the positive and then the negative leads from the battery terminals by slackening the retaining nuts and bolts, or by unscrewing the retaining screws if these are fitted.
ining screws if these are fitted.
2. Remove the battery clamp and carefully lift the battery out of its compartment. Hold the battery vertical to ensure that none of the electrolyte is spilled.
3. Replacement is a direct reversal of this procedure. Smear the terminals with petroleum jelly (vaseline) to prevent corrosion. NEVER use an ordinary grease as applied to other parts of the car.

3. BATTERY - MAINTENANCE & INSPECTION

1. Normal weekly battery maintenance consists of checking the electrolyte level of each cell to ensure that the separators are covered by 1/4 in. of electrolyte. If the level has fallen top up the battery using distilled water only. Do not overfill. If a battery is overfilled or any electrolyte spilled, immediately wipe away the excess as electrolyte attacks and corrodes any metal it comes into contact with very rapidly.
2. As well as keeping the terminals clean and covered with petroleum jelly, the top of the battery, and especially the top of the cells, should be kept clean and dry. This helps prevent corrosion and ensures that the battery does not become partially discharged by leakage through dampness and dirt.
3. Once every three months remove the battery and inspect the battery securing bolts, the battery clamp plate, tray, and battery leads for corrosion (white fluffy deposits on the metal which are brittle to touch). If any corrosion is found, clean off the deposits with ammonia and paint over the clean metal with an anti-rust/anti-acid paint.
4. At the same time inspect the battery case for cracks. If a crack is found, clean and plug it with one of the proprietary compounds marketed by firms such as Holts for this purpose. If leakage through the crack has been excessive then it will be necessary to refill the appropriate cell with fresh electrolyte as detailed later. Cracks are frequently caused to the top of the battery cases by pouring in distilled water in the middle of winter AFTER instead of BEFORE a run. This gives the water no chance to mix with the electrolyte and so the former freezes and splits the battery case.
5. If topping up the battery becomes excessive and the case has been inspected for cracks that could cause leakage, but none are found, the battery is being overcharged and the voltage regulator will have to be checked and reset.
6. With the battery on the bench at the three monthly interval check, measure its specific gravity with a hydrometer to determine the state of charge and condition of the electrolyte. There should be very little variation between the different cells and if a variation in excess of 0.025 is present it will be due to either:

a) Loss of electrolyte from the battery at some time caused by spillage or a leak resulting in a drop in the specific gravity of the electrolyte, when the deficiency was replaced with distilled water instead of fresh electrolyte.

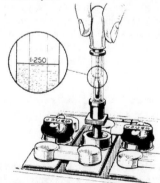

Fig. 10.1. Taking a specific gravity reading with a hydrometer.

b) An internal short circuit caused by buckling of the plates or a similar malady pointing to the likelihood of total battery failure in the near future.

7. The specific gravity of the electrolyte for fully charged conditions at the electrolyte temperature indicated, is listed in Table A. The specific gravity of a fully discharged battery at different temperatures of the electrolyte is given in Table B.

8. Specific gravity is measured by drawing up into the body of a hydrometer sufficient electrolyte to allow the indicator to float freely (see Fig. 10.1). The level at which the indicator floats indicates the specific gravity.

TABLE A

Specific Gravity - Battery fully charged

1.268 at 100°F or 38°C electrolyte temperature
1.272 at 90°F or 32°C   "   "
1.276 at 80°F or 27°C   "   "
1.280 at 70°F or 21°C   "   "
1.284 at 60°F or 16°C   "   "
1.288 at 50°F or 10°C   "   "
1.292 at 40°F or 4°C   "   "
1.296 at 30°F or -1.5°C   "   "

TABLE B

Specific Gravity - Battery fully discharged

1.098 at 100°F or 38°C electrolyte temperature
1.102 at 90°F or 32°C   "   "
1.106 at 80°F or 27°C   "   "
1.110 at 70°F or 21°C   "   "
1.114 at 60°F or 16°C   "   "
1.118 at 50°F or 10°C   "   "
1.122 at 40°F or 4°C   "   "
1.126 at 30°F or -1.5°C   "   "

4. ELECTROLYTE REPLENISHMENT

1. If the battery is in a fully charged state and one of the cells maintains a specific gravity reading which is 0.025 or more lower than the others, and a check of each cell has been made with a voltage meter to check for short circuits (a four to seven second test should give a steady reading of between 1.2 to 1.8 volts), then it is likely that electrolyte has been lost from the cell with the low reading at some time.

2. Top the cell up with a solution of 1 part sulphuric acid to 2.5 parts of water. If the cell is already fully topped up draw some electrolyte out of it with a pipette. The total capacity of each cell is $^3\!/_4$ pint.

3. When mixing the sulphuric acid and water NEVER ADD WATER TO SULPHURIC ACID - always pour the acid slowly onto the water in a glass container. IF WATER IS ADDED TO SULPHURIC ACID IT WILL EXPLODE.

4. Continue to top up the cell with the freshly made electrolyte and then recharge the battery and check the hydrometer readings.

5. BATTERY CHARGING

1. In winter time when heavy demand is placed upon the battery, such as when starting from cold, and much electrical equipment is continually in use, it is a good idea to occasionally have the battery fully charged from an external source at the rate of 3.5 to 4 amps.

2. Continue to charge the battery at this rate until no further rise in specific gravity is noted over a four hour period.

3. Alternatively, a trickle charger charging at the rate of 1.5 amps can be safely used overnight.

4. Specially rapid 'boost' charges which are claimed to restore the power of the battery in 1 to 2 hours are most dangerous as they can cause serious damage to the battery plates through over-heating.

5. While charging the battery note that the temperature of the electrolyte should never exceed 100°F.

6. DYNAMO - ROUTINE MAINTENANCE

1. Routine maintenance consists of checking the tension of the fan belt, and lubricating the dynamo rear bearing once every 12,000 miles.

2. The fan belt should be tight enough to ensure no slip between the belt and the dynamo pulley. If a shrieking noise comes from the engine when the unit is accelerated rapidly, it is likely that it is the fan belt slipping. On the other hand, the belt must not be too taut or the bearings will wear rapidly and cause dynamo failure or bearing seizure. Ideally ½ in. of total free movement should be available at the fan belt midway between the fan and the dynamo pulley.

3. To adjust the fan belt tension slightly slacken the three dynamo retaining bolts, and swing the dynamo on the upper two bolts outwards to increase the tension, and inwards to lower it.

4. It is best to leave the bolts fairly tight so that considerable effort has to be used to move the dynamo; otherwise it is difficult to get the correct setting. If the dynamo is being moved outwards to increase the tension and the bolts have only been slackened a little, a long spanner acting as a lever placed behind the dynamo with the lower end resting against the block works very well in moving the dynamo outwards. Retighten the dynamo bolts and check that the dynamo pulley is correctly aligned with the fan belt.

5. Lubrication of the dynamo consists of inserting three drops of S.A.E. 30 engine oil in the small oil hole in the centre of the commutator end bracket. This lubricates the rear bearing. The front bearing is pre-packed with grease and requires no attention.

7. DYNAMO - TESTING IN POSITION

1. If, with the engine running no charge comes from the dynamo, or the charge is very low, first check that the fan belt is in place and is not slipping. Then check that the leads from the control box to the dynamo are firmly attached and that one has not come loose from its terminal.

2. The lead from the larger 'D' terminal on the dynamo should be connected to the 'D' terminal on the control box, and similarly the 'F' terminals on the dynamo and control box should also be connected together. Check that this is so and that the leads have not been incorrectly fitted. Ensure that a good connection exists to control box terminal 'E'.

3. Make sure none of the electrical equipment (such as the lights or radio) is on and then pull the leads off the dynamo terminals marked 'D' and 'F', join the terminals together with a short length of wire.

4. Attach to the centre of this length of wire the

negative clip of a 0-20 volts voltmeter and run the other clip to earth on the dynamo yoke. Start the engine and allow it to idle at approximately 750 r.p.m. At this speed the dynamo should give a reading of about 15 volts on the voltmeter. There is no point in raising the engine speed above a fast idle as the reading will then be inaccurate.

5. If no reading is recorded then check the brushes and brush connections. If a very low reading of approximately 1 volt is observed then the field winding may be suspect.

6. If a reading of between 4 to 6 amps is recorded it is likely that the armature winding is at fault.

7. If the voltmeter shows a good reading then with the temporary link still in position connect both leads from the control box to 'D' and 'F' on the dynamo ('D' to 'D' and 'F' to 'F'). Release the lead from the 'D' terminal at the control box end and clip one lead from the voltmeter to the end of the cable, and the other lead to a good earth. With the engine running at the same speed as previously, an identical voltage to that recorded at the dynamo should be noted on the voltmeter. If no voltage is recorded then there is a break in the wire. If the voltage is the same as recorded at the dynamo then check the 'F' lead in similar fashion. If both readings are the same as at the dynamo then it will be necessary to test the control box.

8. **DYNAMO - REMOVAL & REPLACEMENT**

1. Slacken the two dynamo retaining bolts, and the nut on the sliding link, and move the dynamo in towards the engine so that the fan belt can be removed.

2. Disconnect the two leads from the dynamo terminals.

3. Remove the nut from the sliding link bolt, and remove the two upper bolts. The dynamo is then free to be lifted away from the engine.

4. Replacement is a reversal of the above procedure. Do not finally tighten the retaining bolt and the nut on the sliding link until the fan belt has been tensioned correctly. See 10/6.2 for details.

9. **DYNAMO - DISMANTLING & INSPECTION**

1. Mount the dynamo in a vice and unscrew and remove the two through bolts from the commutator end bracket. (See photo).

2. Mark the commutator end bracket and the dynamo casing so the end bracket can be replaced in its original position. Pull the end bracket off the armature shaft. NOTE some versions of the dynamo may have a raised pip on the end bracket which locates in a recess on the edge of the casing. If so, marking the end bracket and casing is not necessary. A pip may also be found on the drive end bracket at the opposite end of the casing. (See photo).

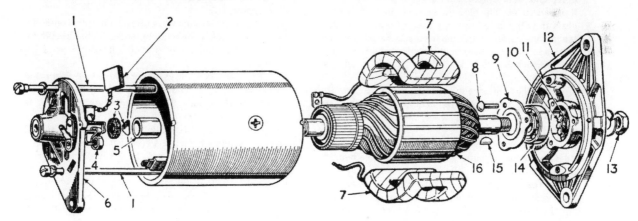

Fig. 10.2. EXPLODED VIEW OF THE C40-L DYNAMO
1 Bolts. 2 Brush. 3 Felt ring and aluminium seating disc. 4 Brush tension spring. 5 Bearing bush. 6 Commutator end bracket. 7 Field coils. 8 Rivet. 9 Bearing retainer plate. 10 Corrugated washer. 11 Felt washer. 12 Driving end bracket. 13 Pulley retainer nut. 14 Bearing. 15 Woodruff key. 16 Armature.

3. Lift the two brush springs and draw the brushes out of the brush holders (arrowed).

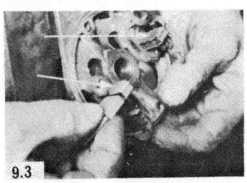

4. Measure the brushes and if worn down to 9/32 in. or less unscrew the screws holding the brush leads to the end bracket. Take off the brushes complete with leads. Old and new brushes are compared in the photograph.

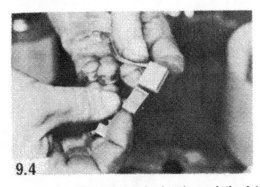

5. If no locating pip can be found, mark the drive end bracket and the dynamo casing so the drive end bracket can be replaced in its original position. Then pull the drive end bracket complete with armature out of the casing.
6. Check the condition of the ball bearing in the drive end plate by firmly holding the plate and noting if there is visible side movement of the armature shaft in relation to the end plate. If play is present the armature assembly must be separated from the end plate. If the bearing is sound there is no need to carry out the work described in the following two paragraphs.
7. Hold the armature in one hand (mount it carefully in a vice if preferred) and undo the nut holding the pulley wheel and fan in place. Pull off the pulley wheel and fan.
8. Next remove the woodruff key (arrowed) from its slot in the armature shaft and also the bearing locating ring.
9. Place the drive end bracket across the open jaws of a vice with the armature downwards and gently tap the armature shaft from the bearing in the end plate with the aid of a suitable drift.
10. Carefully inspect the armature and check it for open or short circuited windings. It is a good indication of an open circuited armature when the commutator segments are burnt. If the armature has short circuited the commutator segments will be very badly burnt, and the overheated armature windings badly discoloured. If open or short circuits are suspected then test by substituting the suspect armature for a new one.
11. Check the resistance of the field coils. To do this, connect an ohmmeter between the field terminals and the yoke and note the reading on the ohmmeter which should be about 6 ohms. If the ohmmeter reading is infinity this indicates an open circuit in the field winding. If the ohmmeter reading is below 5 ohms this indicates that one of the field coils is faulty and must be replaced.
12. Field coil replacement involves the use of a wheel operated screwdriver, a soldering iron, caulking and riveting and this operation is considered to be beyond the scope of most owners. Therefore, if the field coils are at fault either purchase a rebuilt dynamo, or take the casing to a Triumph dealer or electrical engineering works for new field coils to be fitted.
13. Next check the condition of the commutator (arrowed). If it is dirty and blackened as shown, clean it with a petrol dampened rag. If the commutator is in good condition the surface will be smooth and quite free from pits or burnt areas, and the insulated segments clearly defined.
14. If, after the commutator has been cleaned pits and burnt spots are still present, wrap a strip of glass paper round the commutator taking great care to move the commutator 1/4 of a turn every ten rubs till it is thoroughly clean.
15. In extreme cases of wear the commutator can be mounted in a lathe and with the lathe turning at high speed, a very fine cut may be taken off the commutator. Then polish the commutator with glass paper. If the commutator has worn so that the insulators between the segments are level with the top of the segments, then undercut the insulators to a depth of 1/32 in. (.8mm). The best tool to use for this purpose is half a hacksaw blade ground to a thickness of the insulator, and with the handle end of the blade covered in insulating tape to make it comfortable to hold. This is the sort of finish the surface of the commutator should have when finished. (See photo).
16. Check the bush bearing (arrowed) in the commutator end bracket for wear by noting if the armature spindle rocks when placed in it. If worn it must be renewed.
17. The bush bearing can be removed by a suitable extractor or by screwing a 5/8 in. tap four or five times into the bush. The tap complete with bush is then pulled out of the end bracket.
18. NOTE before fitting the new bush bearing that it is of the porous bronze type, and it is essential that it is allowed to stand in S.A.E. 30 engine oil for at least 24 hours before fitment. In an emergency the bush can be immersed in hot oil (100°C) for 2 hours.
19. Carefully fit the new bush into the end plate, pressing it in until the end of the bearing is flush with the inner side of the end plate. If available press the bush in with a smooth shouldered mandrel the same diameter as the armature shaft.

10. DYNAMO - REPAIR & REASSEMBLY
1. To renew the ball bearing fitted to the drive end bracket drill out the rivets which hold the bearing retainer plate to the end bracket and lift off the plate.

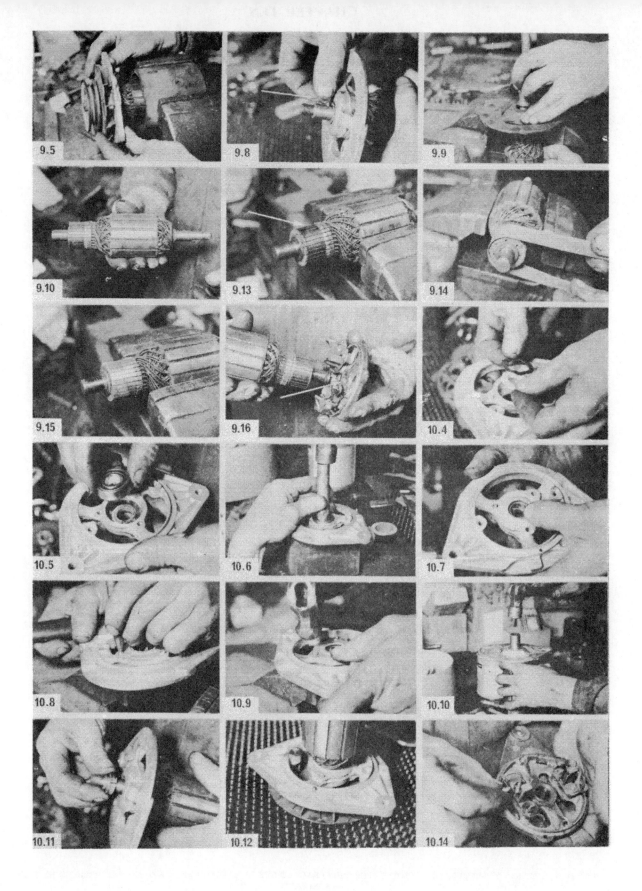

CHAPTER TEN

2. Press out the bearing from the end bracket and remove the corrugated and felt washers from the bearing housing.
3. Thoroughly clean the bearing housing, and the new bearing and pack with high melting-point grease.
4. Place the felt washer and corrugated washer in that order in the end bracket bearing housing.
5. Then fit the new bearing as shown.
6. Gently tap the bearing into place with the aid of a suitable drift.
7. Replace the bearing plate and fit three new rivets.
8. Open up the rivets with the aid of a suitable cold chisel.
9. Finally peen over the open end of the rivets with the aid of a ball hammer as illustrated.
10. Refit the drive end bracket to the armature shaft. Do not try and force the bracket on but with the aid of a suitable socket abutting the bearing tap the bearing in gently, so pulling the end bracket down with it.
11. Slide the spacer up the shaft and refit the woodruff key.
12. Replace the fan and pulley wheel and then fit the spring washer and nut and tighten the latter. The drive bracket end of the dynamo is now fully assembled as shown.
13. If the brushes are little worn and are to be used again then ensure that they are placed in the same holders from which they were removed. When refitting brushes, either new or old, check that they move freely in their holders. If either brush sticks, clean with a petrol moistened rag and if still stiff, lightly polish the sides of the brush with a very fine file until the brush moves quite freely in its holder.
14. Tighten the two retaining screws and washers which hold the wire leads to the brushes in place.
15. It is far easier to slip the end piece with brushes over the commutator if the brushes are raised in their holders as shown and held in this position by the pressure of the springs resting against their flanks (arrowed).
16. Refit the armature to the casing and then the commutator end plate and screw up the two through bolts.

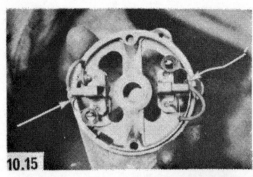

17. Finally, hook the ends of the two springs off the flanks of the brushes and onto their heads so the brushes are forced down into contact with the armature.

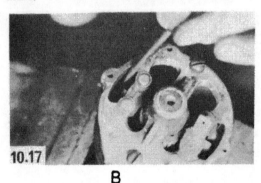

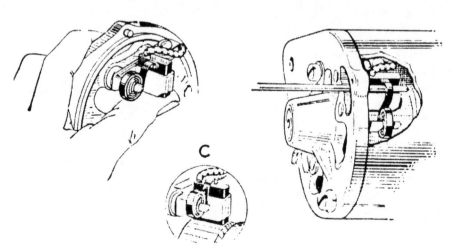

Fig. 10.3. METHOD OF FITTING THE COMMUTATOR END BRACKET AFTER RAISING AND TRAPPING THE BRUSHES BY THEIR SPRINGS
A Brush trapped by spring in raised position. B Releasing the brush onto the commutator. C Normal position of brush.

ELECTRICAL SYSTEM

11. STARTER MOTOR - GENERAL DESCRIPTION

The starter motor is mounted on the left hand side of the engine end plate, and is held in position by two bolts which also clamp the bellhousing flange. The motor is of the four field coil, four pole piece type, and utilises four spring-loaded commutator brushes. Two of these brushes are earthed, and the other two are insulated and attached to the field coil ends.

12. STARTER MOTOR - TESTING IN ENGINE

1. If the starter motor fails to operate then check the condition of the battery by turning on the headlamps. If they glow brightly for several seconds and then gradually dim, the battery is in an uncharged condition.
2. If the headlamps glow brightly and it is obvious that the battery is in good condition then check the tightness of the battery wiring connections (and in particular the earth lead from the battery terminal to its connection on the bodyframe). Check the tightness of the connections at the relay switch and at the starter motor. Check the wiring with a voltmeter for breaks or shorts.
3. If the wiring is in order then check that the starter motor switch is operating. To do this press the rubber covered button in the centre of the relay switch under the bonnet. If it is working the starter motor will be heard to 'click' as it tries to rotate. Alternatively check it with a voltmeter.
4. If the battery is fully charged, the wiring in order, and the switch working and the starter motor fails to operate then it will have to be removed from the car for examination. Before this is done, however, ensure that the starter pinion has not jammed in mesh with the flywheel. Check by turning the square end of armature shaft with a spanner. This will free the pinion if it is stuck in engagement with the flywheel teeth.

13. STARTER MOTOR - REMOVAL & REPLACEMENT

1. Disconnect the battery earth lead from the positive terminal (negative on later models).
2. Disconnect the heavy lead to the starter motor at the solenoid terminal. (This end is much more accessible than the end fitted to the starter motor.).
3. Undo and remove the two bolts which hold the starter motor in place and withdraw it upwards together with the distance piece.
4. Generally replacement is a straightforward reversal of the removal sequence. Check that the electrical cable is firmly attached to the starter motor terminal before fitting the starter motor in place. Also ensure that any packing washers originally fitted are replaced so as to give the correct out of mesh clearance between the stationary starter pinion and the flywheel ring gear of $3/32$ in. to $5/32$ in.

14. STARTER MOTOR - DISMANTLING & REASSEMBLY

1. With the starter motor on the bench, loosen the screw on the cover band and slip the cover band off. With a piece of wire bent into the shape of a hook, lift back each of the brush springs in turn and check the movement of the brushes in their holders by pulling on the flexible connectors. If the brushes are so worn that their faces do not rest against the commutator or if the ends of the brush leads are exposed on their working face, they must be renewed.
2. If any of the brushes tend to stick in their holders then wash them with a petrol moistened cloth and, if necessary, lightly polish the sides of the brush with a very fine file, until the brushes move quite freely in their holders.
3. If the surface of the commutator is dirty or blackened, clean it with a petrol dampened rag. Secure the starter motor in a vice and check it by connecting a heavy gauge cable between the starter motor terminal and a 12-volt battery.
4. Connect the cable from the other battery terminal to earth in the starter motor body. If the motor turns at high speed it is in good order.
5. If the starter motor still fails to function or if it is wished to renew the brushes, then it is necessary

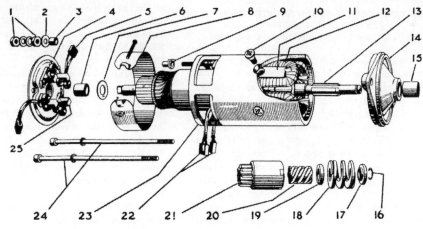

Fig. 10.4. EXPLODED VIEW OF THE STARTER MOTOR

1 Terminal nuts and washers. 2 Insulating washer. 3 Insulating bush. 4 End plates. 5 Brush. 6 Bush. 7 Thrust washer. 8 Cover band. 9 Insulating bush. 10 Pole securing screw. 11 Pole piece. 12 Field coil. 13 Shaft. 14 End bracket. 15 Bush. 16 Jump ring. 17 Retainer. 18 Main spring. 19 Thrust washer. 20 Sleeve. 21 Pinion and barrel assembly. 22 Brushes. 23 Yoke. 24 Through bolts. 25 Brush box.

to further dismantle the motor.

6. Lift the brush springs with the wire hook and lift all four brushes out of their holders one at a time.

7. Remove the terminal nuts and washers from the terminal post on the commutator end bracket.

8. Unscrew the two through bolts which hold the end plates together and pull off the commutator end bracket. Also remove the driving end bracket which will come away complete with the armature.

9. At this stage if the brushes are to be renewed, their flexible connectors must be unsoldered and the connectors of new brushes soldered in their place. Check that the new brushes move freely in their holders as detailed above. If cleaning the commutator with petrol fails to remove all the burnt areas and spots, then wrap a piece of glass paper round the commutator and rotate the armature.

Fig. 10.5. The correct method of undercutting commutator insulation

10. If the commutator is very badly worn, remove the drive gear as detailed in the following section. Then mount the armature in a lathe and with the lathe turning at high speed, take a very fine cut out of the commutator and finish the surface by polishing with glass paper, DO NOT UNDERCUT THE MICA INSULATORS BETWEEN THE COMMUTATOR SEGMENTS.

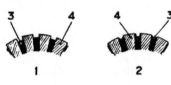

Fig. 10.6. SHOWING CORRECT AND INCORRECT UNDERCUTTING OF THE COMMUTATOR
1 Correct. 2 Incorrect. 3 Insulation. 4 Segments.

11. With the starter motor dismantled, test the four field coils for an open circuit. Connect a 12-volt battery with a 12-volt bulb in one of the leads between the field terminal post and the tapping point of the field coils to which the brushes are connected. An open circuit is proved by the bulb not lighting.

12. If the bulb lights, it does not necessarily mean that the field coils are in order, as there is a possibility that one of the coils will be earthed to the starter yoke or pole shoes. To check this, remove the lead from the brush connector and place it against a clean portion of the starter yoke. If the bulb lights the field coils are earthing. Replacement of the field coils calls for the use of a wheel operated screwdriver, a soldering iron, caulking and riveting operations and is beyond the scope of the majority of owners. The starter yoke should be taken to a reputable electrical engineering works for new field coils to be fitted. Alternatively, purchase an exchange Lucas starter motor.

13. If the armature is damaged this will be evident after visual inspection. Look for signs of burning, discolouration, and for conductors that have lifted away from the commutator. Reassembly is a straightforward reversal of the dismantling procedure.

15. STARTER MOTOR DRIVE - GENERAL DESCRIPTION

1. The starter motor drive is of the outboard type. When the starter motor is operated the pinion moves into contact with the flywheel gear ring by moving in towards the starter motor.

2. If the engine kicks back, or the pinion fails to engage with the flywheel gear ring when the starter motor is actuated no undue strain is placed on the armature shaft, as the pinion sleeve disengages from the pinion and turns independently.

16. STARTER MOTOR DRIVE - REMOVAL & REPLACEMENT

1. Either one of two types of starter motor drive may be fitted. Early cars make use of the drive shown in Fig. 10.7. To dismantle the earlier drive, extract the split pin from the shaft nut on the end of the starter drive.

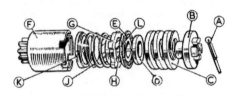

Fig. 10.7. EXPLODED VIEW OF THE STARTER DRIVE FITTED TO EARLIER MODELS
A Split pin. B Shaft nut. C Main spring. D Buffer washer.
E Retaining spring. F Pinion and barrel. G Control nut. H Sleeve
J Retaining spring. K Splined washer. L Corrugated washer.

2. Holding the squared end of the armature shaft at the commutator end bracket with a suitable spanner, unscrew the shaft nut which has a right-hand thread, and pull off the mainspring.

3. Slide the remaining parts with a rotary action off the armature shaft.

4. Reassembly is a straightforward reversal of the above procedure. Ensure that the split pin is refitted. NOTE: It is most important that the drive gear is completely free from oil, grease and dirt. With the drive gear removed, clean all the parts thoroughly in paraffin. UNDER NO CIRCUMSTANCES OIL THE DRIVE COMPONENTS. Lubrication of the drive components could easily cause the pinion to stick.

5. The later type of drive shown in Fig. 10.4 has a circlip at the retaining end of the shaft instead of a nut and split pin.

6. To dismantle the later drive use a press to push the retainer clear of the circlip which can then be removed. The remainder of the dismantling sequence is the same as for the earlier type. On replacement use a suitable press to compress the spring and retainer sufficiently to allow a new circlip to be fitted to its groove on the shaft. Remove the press.

ELECTRICAL SYSTEM

17. STARTER MOTOR BUSHES - INSPECTION, REMOVAL & REPLACEMENT

1. With the starter motor stripped down check the condition of the bushes. They should be renewed when they are sufficiently worn to allow visible side movement of the armature shaft.
2. The old bushes are simply driven out with a suitable drift and the new bushes inserted by the same method. As the bearings are of the phospher bronze type it is essential that they are allowed to stand in S.A.E. 30 engine oil for at least 24 hours before fitment.

18. CONTROL BOX - GENERAL DESCRIPTION

The control box comprises the voltage regulator and the cut-out. The voltage regulator controls the output from the dynamo depending on the state of the battery and the demands of the electrical equipment and ensures that the battery is not overcharged. The cut-out is really an automatic switch and connects the dynamo to the battery when the dynamo is turning fast enough to produce a charge. Similarly it disconnects the battery from the dynamo when the engine is idling or stationary so that the battery does not discharge through the dynamo.

19. CUT-OUT & REGULATOR CONTACTS - MAINTENANCE

1. Every 12,000 miles check the cut-out and regulator contacts. If they are dirty or rough or burnt, place a piece of fine glass paper (DO NOT USE EMERY PAPER OR CARBORUNDUM PAPER) between the cut-out contacts, close them manually and draw the glass paper through several times.
2. Clean the regulator contacts in exactly the same way, but use emery or carborundum paper and not glass paper. Carefully clean both sets of contacts from all traces of dust with a rag moistened in methylated spirits.

20. VOLTAGE REGULATOR ADJUSTMENT

1. If the battery is in sound condition, but is not holding its charge, or is being continually overcharged and the dynamo is in sound condition, then the voltage regulator in the control box must be adjusted.

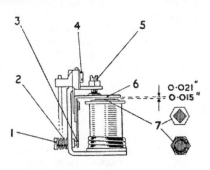

Fig. 10.8. THE AIR GAP SETTINGS ON THE REGULATOR
1 Voltage adjusting screw. 2 Spring. 3 Armature tension spring. 4 Armature securing screws. 5 Fixed contact adjusting screw. 6 Armature. 7 Core face and shim.

2. Check the regulator setting by removing and joining together the cables from the control box terminals A1 and A. Then connect the negative lead of a 20-volt voltmeter to the 'D' terminal on the dynamo and the positive lead to a good earth. Start the engine and increase its speed until the voltmeter needle flicks and then steadies. This should occur at about 2,000 r.p.m. If the voltage at which the needle steadies is outside the limits listed below, then remove the control box cover and turn the adjusting screw, (1) in Fig. 10.8, clockwise a quarter of a turn at a time to raise the setting and a similar amount, anti-clockwise, to lower it.

Air Temperature	Type RB 106/2 Open circuit voltage
10°C or 50°F	16.1 to 16.7
20°C or 68°F	16.0 to 16.6
30°C or 86°F	15.9 to 16.5
40°C or 104°F	15.8 to 16.4

3. It is vital that the adjustments be completed within 30 seconds of starting the engine as otherwise the heat from the shunt coil will affect the readings.

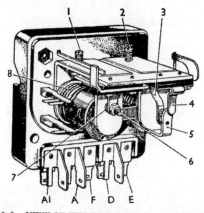

Fig. 10.9. VIEW OF THE REGULATOR & CONTROL BOX
1 Regulator adjusting screw. 2 Cut-out adjusting screw. 3 Fixed contact blade. 4 Stop arm. 5 Armature tongue and moving contact. 6 Regulator fixed contact screw. 7 Regulator moving contact. 8 Windings.

21. CUT-OUT ADJUSTMENT

1. Check the voltage required to operate the cut-out by connecting a voltmeter between the control box terminals 'D' and 'E'.
2. Remove the control box cover, start the engine and gradually increase its speed until the cut-outs close. This should occur when the reading is between 12.7 to 13.3 volts.
3. If the reading is outside these limits turn the cut-out adjusting screw, (2) in Fig. 10.9 a fraction at a time clockwise to raise the voltage, and anti-clockwise to lower it. To adjust the drop off voltage bend the fixed contact blade carefully. The adjustment to the cut-out should be completed within 30 seconds of starting the engine as otherwise heat build-up from the shunt coil will affect the readings.
4. If the cut-out fails to work, clean the contacts, and, if there is still no response, renew the cut-out and regulator unit.

CHAPTER TEN

22. FLASHER CIRCUIT - FAULT TRACING & RECTIFICATION

1. The actual flasher unit consists of a small alloy cylindrical container positioned either under the fascia or on the rear bulkhead of the engine compartment.
2. If the flasher unit works twice as fast as usual when indicating either right or left this is a sure indication of a broken filament in the front or rear indicator bulb on the side operating too quickly.
3. If the external flashers are working but the internal flasher warning light has ceased to function check the filament of the warning bulb and replace as necessary.
4. With the aid of the wiring diagram check all the flasher circuit connections if a flasher bulb is sound but does not work.
5. With the ignition turned on check that current is reaching the flasher unit by connecting a voltmeter between the 'plus' or 'B' terminal and earth. If this test is positive connect the 'plus' or 'B' terminal and the 'L' terminal and operate the flasher switch. If the flasher bulb lights up the flasher unit itself is defective and must be replaced as it is not possible to dismantle and repair it.

23. WINDSCREEN WIPER MECHANISM - MAINTENANCE

1. Renew the windscreen wiper blades at intervals of 12,000 miles, or more frequently if necessary.
2. The cable which drives the wiper blades from the gearbox attached to the windscreen wiper motor is pre-packed with grease and requires no maintenance. The washer round the wheelbox spindle can be lubricated with several drops of glycerine every 6,000 miles.

24. WINDSCREEN WIPER ARMS - REMOVAL & REPLACEMENT

1. Before removing a wiper arm, turn the windscreen wiper switch on and off to ensure the arms are in their normal parked position parallel with the bottom of the windscreen.
2. To remove an arm pivot the arm back and pull the wiper arm head off the splined drive.
3. When replacing an arm position it so it is in the correct relative parked position and then press the arm head onto the splined drive till the retaining clip clicks into place.

25. WINDSCREEN WIPER MECHANISM - FAULT DIAGNOSIS & RECTIFICATION

1. Should the windscreen wipers fail to park or park badly then check the limit switch on the gearbox cover.
2. Loosen the four screws which retain the gearbox cover and place the projection close to the rim of the limit switch in line with the groove in the gearbox cover.
3. Rotate the limit switch anti-clockwise 25° and tighten the four screws retaining the gearbox cover. If it is wished to park the windscreen wipers on the other side of the windscreen rotate the limit switch 180° clockwise.
4. Should the windscreen wipers fail, or work very slowly, then check the current the motor is taking by connecting up a 1-20 volt voltmeter in the circuit and turning on the wiper switch. Consumption should be between 2.7 to 3.4 amps.
5. If no current is passing through check the wiring for breaks and loose connections.
6. If the wiper motor takes a very high current check the wiper blades for freedom of movement. If this is satisfactory check the gearbox cover and gear assembly for damage and measure the armature endfloat which should be between .008 to .012 in. (.20 to .30 mm.).
7. The endfloat is set by the adjusting screw. Check that excessive friction in the cable connecting tubes caused by too small a curvature is not the cause of the high current consumption.
8. If the motor takes a very low current ensure that the battery is fully charged. Check the brush gear after removing the commutator end bracket or cover and ensure that the brushes are bearing on the commutator. If not, check the brushes for freedom of movement and if necessary renew the tension spring.
9. Check the armature by substitution if this unit is suspected.

26. WIPER MOTOR - REMOVAL & REPLACEMENT

1. The wiper motor is always removed in unit with the flexible inner cable rack (18 in Fig. 10.12). First pull off the wiper arms and blades.
2. Unscrew the large nut which holds the rigid tubing (19) to the wiper motor gearbox.
3. Pull off the three wires from their Lucar connectors on the end cover (12).
4. Undo the nuts and washers from the three bolts which hold the wiper motor gearbox in place and take off the motor complete with the inner cable rack. On some models it will be necessary to undo the bolts which hold the motor mounting bracket to the dash panel before access can be gained to the nuts and washers which hold the motor to the mounting bracket.
5. Replacement is a straightforward reversal of the removal sequence. Lubricate the flexible inner cable with Ragosine Listate grease or similar to ensure smooth functioning in the rigid tubing.

27. WHEELBOX - REMOVAL & REPLACEMENT

1. Remove the wiper motor complete with gearbox and flexible inner cable rack as described in the previous section.
2. The windscreen wiper arm wheelboxes are located immediately underneath the splined drive shafts over which the wiper arms fit. To remove these wheelboxes release the cable rack outer casings by slackening the wheelbox cover screws. Remove the external nut, bush, and washer from the base of the splines and pull out the wheelboxes from under the fascia.
3. Replacement is a straight reversal of the above sequence but take care that the cable rack emerges properly and that the wheelboxes are correctly lined up.

28. WINDSCREEN WIPER MOTOR - DISMANTLING, INSPECTION & REASSEMBLY

1. Mark the domed cover in relation to the flat gearbox lid, undo the four screws holding the gearbox lid in place and lift off the lid and domed cover.

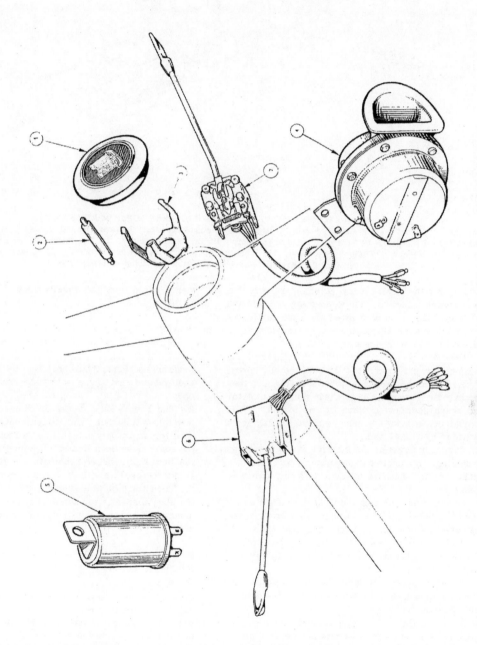

Fig. 10.10. EXPLODED VIEW OF HORN, FLASHER AND LIGHT CONTROLS

1 Horn push assembly. 2 Horn push connection brush. 3 Horn push assembly clip. 4 Windtone horn assembly (pair). 5 Flasher unit. 6 Flasher indicator switch. 7 Lighting switch—side, head and dip.

CHAPTER TEN

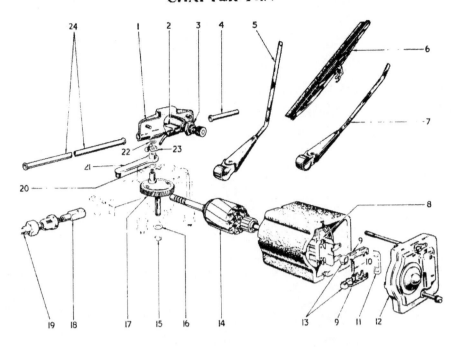

Fig. 10.11. EXPLODED VIEW OF WINDSCREEN WIPER MECHANISM
1 Wheel box. 2 Jet and bush assembly. 3 Nut. 4 Rigid tubing—right-hand side. 5 Wiper arm. 6 Blade. 7 Wiper arm. 8 Field coil assembly. 9 Brushgear. 10 Tension spring and retainers. 11 Brushgear retainer. 12 End cover. 13 Brushes. 14 Armature. 15 Circlip. 16 Washer. 17 Final drive wheel. 18 Cable rack. 19 Rigid tubing—left-hand side. 20 Spacer. 21 Connecting rod. 22 Circlip. 23 Parking switch contact. 24 Rigid tubing—centre section.

2. Pull off the small circlip (22), and remove the limit switch wiper (23). The connecting rod (21) and cable rack (18) can now be lifted off. Take particular note of the spacer (20) located between the final drive wheel (17) and the connecting rod (21).

3. Undo and remove the two through bolts from the commutator end cover (12). Pull off the end cover.

4. Lift out the brushgear retainer (11) and then remove the brushgear (9). Clean the commutator and brush gear and if worn fit new brushes. The resistance between adjacent commutator segments should be .34 to .41 ohm.

5. Carefully examine the internal wiring for signs of chafing, breaks or charring which would lead to a short circuit. Insulate or replace any damaged wiring.

6. Measure the value of the field resistance which should be between 12.8 to 14 ohms. If a lower reading than this is obtained it is likely that there is a short circuit and a new field coil should be fitted.

7. Renew the gearbox gear teeth if they are damaged, chipped or worn.

8. Reassembly is a straightforward reversal of the dismantling sequence, but ensure the following items are lubricated:-
a) Immerse the self aligning armature bearing in S.A.E. 20 engine oil for 24 hours before assembly.
b) Oil the armature bearings in S.A.E. 20 engine oil.
c) Soak the felt lubricator in the gearbox with S.A.E. 20 engine oil.
d) Grease generously the worm wheel bearings, cross head, guide channel, connecting rod, crankpin,

worm, cable rack and wheelboxes and the final gear shaft.

29. HORNS - FAULT TRACING & RECTIFICATION

1. If a horn works badly or fails completely, first check the wiring leading to it for short circuits and loose connections. Also check that the horn is firmly secured and that there is nothing lying on the horn body.

2. The horn should never be dismantled but it is possible to adjust it. This adjustment is to compensate for wear only and will not affect the tone. At the rear of the horn is a small adjustment screw on the broad rim, nearly opposite the two terminals. Do not confuse this with the large screw in the centre.

3. Turn the adjustment screw anti-clockwise until the horn just fails to sound. Then turn the screw a quarter of a turn clockwise, which is the optimum setting.

30. FUEL CONTENTS GAUGE - FAULT TRACING & RECTIFICATION

1. The fuel contents gauge can mis-read in four different ways. There can be no reading at all; there can be intermittant reading; the needle on the gauge can read too high or too low. The main components of the system comprise a fuel contents gauge, a tank sender unit, and a voltage stabiliser unit. Unfortunately, if any of these items become defective they cannot be repaired but must be exchanged for new units.

2. If there is no reading at all check for loose electrical connections. Providing the wiring is in order

ELECTRICAL SYSTEM

check the components one at a time by substitution.

31. FUEL GAUGE SENDER UNIT - REMOVAL & REPLACEMENT

1. Disconnect the earth lead from the battery and the two wires from their Lucar blade terminals on the sender unit.
2. Undo the six screws which hold the sender unit to the tank, or on models where the sender unit is held in place with a retaining ring use a screwdriver to turn the ring anti-clockwise.
3. Carefully lift the complete unit away making sure that the float lever is not bent or damaged in the process.
4. Replacement of the unit is a reversal of the above process. To ensure a fuel tight joint scrape both the tank and sender gauge mating flanges clean, and always use a new joint gasket, together with sealing compound in the case of the screw retained unit.

32. TEMPERATURE GAUGE — FAULT TRACING & RECTIFICATION

1. If the temperature gauge fails to work or works incorrectly, it will be necessary to test for circuit continuity using an Ohmmeter. Alternatively units can be tested by substitution.
2. The system comprises a temperature transmitter located in the thermostat housing, a voltage stabiliser and the temperature gauge itself.
3. None of these items can be repaired and if any are defective they must be exchanged for replacement units.

33. HEADLAMP BULBS - REMOVAL & REPLACEMENT - FOCUSING

1. Loosen the securing screw (14), (see Fig. 10.12) on models fitted with the peaked rim (15) and pull off the rim and the sealing rubber (13).
2. On other models take off the snap on ring (12) with the assistance of the special tool (supplied in the tool kit), which is inserted behind the lower edge of the rim and then levered sideways. Remove the sealing rubber (11).
3. Press the headlamp in towards the wing with the palms of your hands, at the same time turning the unit slightly anti-clockwise.
4. In this way the large holes in the headlamp rim are rotated so they lie under the heads of the three securing/focusing screws. As the holes are larger than the heads of the screws the headlamp can be pulled off.
5. To replace the bulb twist and pull out the bulb holder from the back of the lamp and remove the bulb which is located by a projection and slot to ensure it is always fitted the correct way round. Replacement is a simple reversal of this process.
6. Do not turn any of the screws especially the three screws with springs under their heads, as these are the focusing screws, one at the top of the headlamp rim for vertical adjustment and one each side of the rim for adjustment to the left and right.
7. In the author's experience the best method of focusing the headlights is to remove the cowls at home and then set the lights up at night to give the best results on a long straight flat road. Always lower the beam of the lights a little from their optimum position so as not to dazzle other road users and to compensate for those occasions when there is weight in the back.
8. Bear in mind that MOT regulations demand that when on main beam the beams must be parallel with the road and with each other.

34. FRONT SIDE & FLASHER BULBS - REMOVAL & REPLACEMENT

1. One bulb with two filaments serves for both the parking/side and flasher lights. If either filament becomes broken the bulb must be replaced.
2. Access to the bulb is simply gained by undoing two screws and lifting off the sidelight rim and lens.

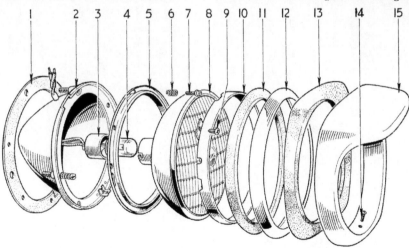

Fig. 10.12. EXPLODED VIEW OF THE TRIUMPH HEADLIGHT COMPONENTS

1 Rubber seal. 2 Housing. 3 Adaptor. 4 Bulb. 5 Inner rim. 6 Spring. 7 Screw. 8 Light unit. 9 Screw. 10 Outer rim. 11 Sealing rubber.* 12 Snap on rim.* 13 Sealing rubber.** 14 Screw.** 15 Rim.**

*These units are used on some models.   **These components are used where the former units are not.

CHAPTER TEN

35. STOP/REAR & FLASHER BULBS - REMOVAL & REPLACEMENT

1. Undo the two screws holding the appropriate red/amber rear lens in place. The flasher bulb is the top bulb with the single filament. The lower twin filament bulb is for the stop/rear lights.
2. Replace the failed bulb and reposition the lens.

36. INSTRUMENT PANEL & WARNING BULBS - REMOVAL & REPLACEMENT

1. The bulb for the direction indicator is removed from behind the fascia. The main beam, ignition and oil warning bulbs, together with the illumination bulbs can all be removed by a simple twist and pull action from the rear of the instrument.
2. The fascia illuminating bulb is easily replaced from the front of the fascia.

37. NUMBER PLATE BULB - REMOVAL & REPLACEMENT

1. Undo the screw in the centre of the chromium cover and lift off the cover and the glass lens.
2. Push, twist and then pull out the old bulb and fit a replacement.

38. FUSE - HERALD 13/60

1. A single 25 amp. fuse is fitted in line to protect the headlamp flasher circuit. The fuse holder is located adjacent to the coil on the bulkhead and the two halves of the holder are held together by a bayonet type fitting.
2. If a replacement fuse is fitted and it too fails, check the system for short circuits or other faults.

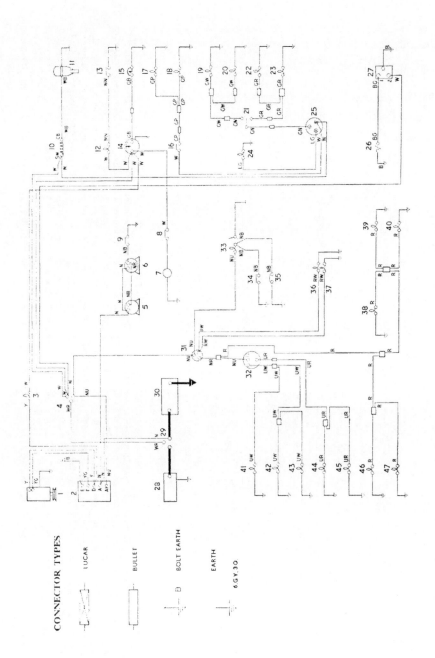

WIRING DIAGRAM - 948 c.c. SINGLE CARBURETTER MODELS

1 Generator. 2 Control box. 3 Ignition warning light. 4 Ignition/start switch. 5 Horn. 6 Horn. 7 Heater motor. 8 Heater switch. 9 Horn push. 10 Ignition coil. 11 Distributor. 12 Oil pressure warning light. 13 Oil pressure switch. 14 Fuel gauge. 15 Stop lamp switch. 16 Fuel tank unit. 17 R.H. stop lamp. 18 L.H. stop lamp. 19 R.H. rear flasher. 20 R.H. front flasher. 21 Flasher switch. 22 L.H. front flasher. 23 L.H. rear flasher. 24 Flasher warning light. 25 Flasher unit. 26 Screen wiper switch. 27 Screen wiper motor. 28 Starter motor. 29 Starter solenoid switch. 30 12 volt battery. 31 Master lighting switch. 32 Lighting switch. 33 Interior light. 34 R.H. courtesy light switch. 35 L.H. courtesy light switch. 36 Panel illumination bulb. 37 Panel illumination bulb. 38 Number plate lamp. 39 R.H. tail lamp. 40 L.H. tail lamp. 41 Main beam warning light. 42 R.H. headlamp main beam. 43 L.H. headlamp main beam. 44 R.H. headlamp dip beam. 45 L.H. headlamp dip beam. 46 L.H. side lamp. 47 R.H. side lamp.

COLOUR CODE

B Black. U Blue. N Brown. G Green. K Pink. P Purple. R Red. S Slate. W White. Y Yellow. D Dark. L Light. M Medium.

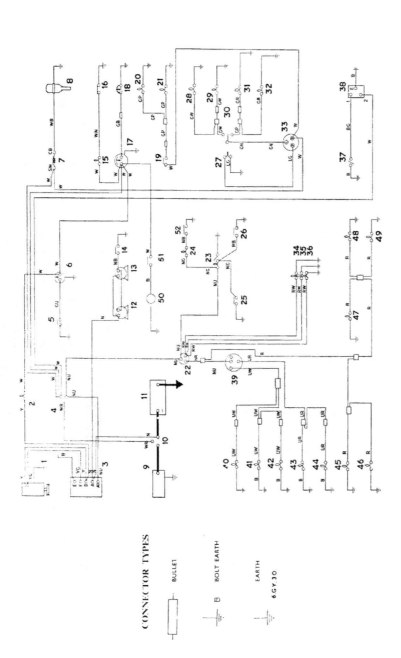

WIRING DIAGRAM - 948 c.c. TWIN CARBURETTER MODELS

1 Dynamo. 2 Ignition warning light. 3 Control box. 4 Ignition—starter switch. 5 Temperature gauge transmitter. 6 Temperature gauge. 7 Ignition coil. 8 Distributor. 9 Starter motor. 10 Starter solenoid switch. 11 Battery. 12 Horn. 13 Horn. 14 Horn switch. 15 Oil pressure warning light. 16 Oil pressure warning light switch. 17 Fuel gauge. 18 Fuel tank unit. 19 Stop lamp switch. 20 R.H. stop lamp. 21 L.H. stop lamp. 22 Lamps master switch. 23 Interior light. 24 Glove locker switch. 25 L.H. courtesy switch. 26 R.H. courtesy switch. 27 Flasher warning light. 28 R.H. rear flasher. 29 R.H. front flasher. 30 Direction indicator switch. 31 L.H. front flasher. 32 L.H. rear flasher. 33 Flasher unit. 34 Speedometer illumination bulb. 35 Temperature gauge illumination bulb. 36 Fuel gauge illumination bulb. 37 Screen wiper switch. 38 Screen wiper motor. 39 Lighting switch (dip and side.) 40 Main beam warning light. 41 R.H. headlamp main beam. 42 L.H. headlamp main beam. 43 R.H. headlamp dip beam. 44 L.H. headlamp dip beam. 45 L.H. side lamp. 46 R.H. side lamp. 47 Number plate lamp. 48 R.H. tail lamp. 49 L.H. tail lamp. 50 Blower motor. 51 Heater switch. 52 Glove locker light switch.

COLOUR CODE

B Black. U Blue. N Brown. G Green. K Pink. P Purple. R Red. S Slate. W White. Y Yellow. D Dark. L Light. M Medium.

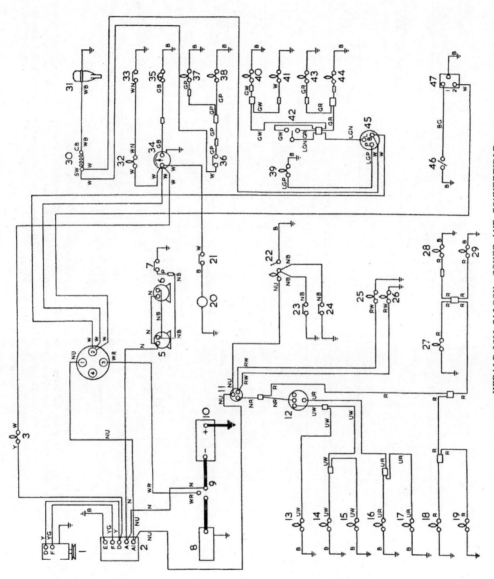

HERALD 1200 SALOON, COUPE AND CONVERTIBLE

1 Generator. 2 Control box. 3 Ignition warning light. 4 Ignition/starter switch. 5 Horn. 6 Horn. 7 Horn push. 8 Starter motor. 9 Starter solenoid switch. 10 Battery. 11 Master lighting switch. 12 Column switch. 13 Main beam warning light. 14 R.H. headlamp main beam. 15 L.H. headlamp main beam. 16 R.H. headlamp dip beam. 17 L.H. headlamp dip beam. 18 L.H. side lamp. 19 R.H. side lamp. 20 Heater motor. 21 Heater switch. 22 Heater switch. 23 R.H. courtesy light switch. 24 L.H. courtesy light switch. 25 Panel illumination. 26 Panel illumination. 27 Number plate lamp. 28 R.H. tail lamp. 29 L.H. tail lamp. 30 Ignition coil. 31 Distributor. 32 Oil pressure warning light. 33 Oil pressure switch. 34 Fuel gauge. 35 Fuel tank unit. 36 Stop lamp switch. 37 R.H. stop lamp. 38 L.H. stop lamp. 39 Flasher warning light. 40 R.H. rear flasher. 41 R.H. front flasher. 42 Flasher switch. 43 L.H. front flasher. 44 L.H. rear flasher. 45 Flasher unit. 46 Screen wiper switch. 47 Screen wiper motor.

161

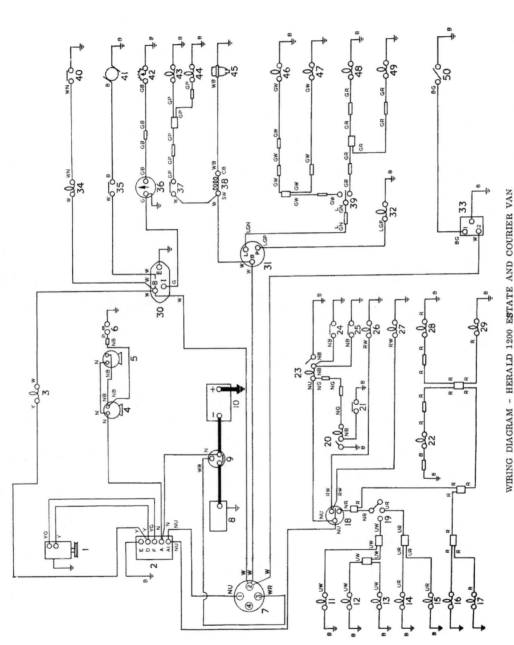

WIRING DIAGRAM - HERALD 1200 ESTATE AND COURIER VAN

1 Dynamo. 2 Control box. 3 Ignition warning light. 4 Horn. 5 Horn. 6 Horn push. 7 Ignition—starter switch. 8 Starter motor. 9 Starter solenoid. 10 Battery. 11 Main beam warning light. 12 R.H. headlamp main beam. 13 L.H. headlamp main beam. 14 R.H. headlamp dip beam. 15 L.H. headlamp dip beam. 16 L.H. side lamp. 17 R.H. sidelamp. 18 Master light switch. 19 Column light switch. 20 Tailgate light and switch. 21 Tailgate switch. 22 Number plate lamp. 23 Interior light and switch. 24 R.H. courtesy light switch. 25 L.H. courtesy light switch. 26 Panel illumination. 27 Panel illumination. 28 R.H. tail lamp. 29 L.H. tail lamp. 30 Voltage stabiliser. 31 Flasher unit. 32 Flasher warning light. 33 Heater switch. 34 Oil pressure warning light. 35 Heater switch. 36 Fuel gauge. 37 Stop lamp switch. 38 Ignition coil. 39 Flasher switch. 40 Oil pressure switch. 41 Heater motor. 42 Tank unit. 43 L.H. stop light. 44 R.H. stop light. 45 Distributor. 46 R.H. rear flasher. 47 R.H. front flasher. 48 L.H. rear flasher. 49 L.H. front flasher. 50 Wiper switch.

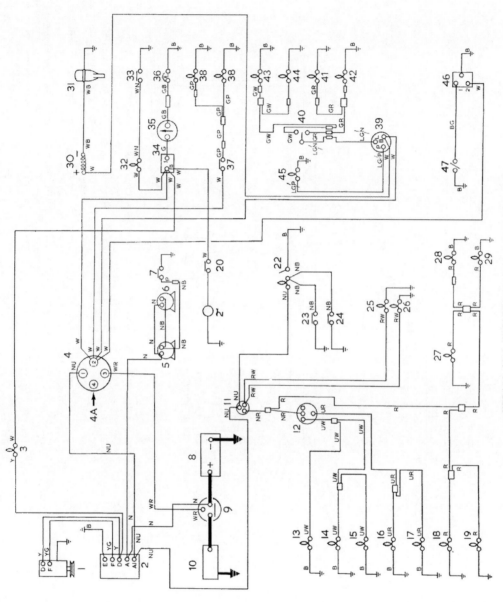

HERALD 1200 NEGATIVE EARTH WIRING DIAGRAM

1 Generator. 2 Control box. 3 Ignition warning light. 4 Ignition/starter switch. 4A Ignition/starter switch radio supply connector. 5 Horn. 6 Horn. 7 Horn push. 8 Battery. 9 Starter solenoid. 10 Starter motor. 11 Master light switch. 12 Column light switch. 13 Main beam warning light. 14 Main beam. 15 Main beam. 16 Dip beam. 17 Dip beam. 18 Front parking lamp. 19 Front parking lamp. 20 Heater switch. 21 Heater motor. 22 Fascia lamp. 23 Door switch. 24 Door switch. 25 Instrument illumination. 26 Instrument illumination. 27 Plate illumination lamp. 28 Tail lamp. 29 Tail lamp. 30 Ignition coil. 31 Ignition distributor. 32 Oil pressure warning light. 33 Oil pressure switch. 34 Voltage stabiliser. 35 Fuel indicator. 36 Fuel tank unit. 37 Stop lamp switch. 38 Stop lamp. 39 Flasher unit. 40 Flasher switch. 41 L.H. flasher lamp. 42 L.H. flasher lamp. 43 R.H. flasher lamp. 44 Flasher lamp. 45 Flasher warning light. 46 Windscreen wiper motor. 47 Windscreen wiper switch.

COLOUR CODE

N Brown. U Blue. R Red. P Purple. G Green. LG Light Green. W White. Y Yellow. S Slate. B Black.

163

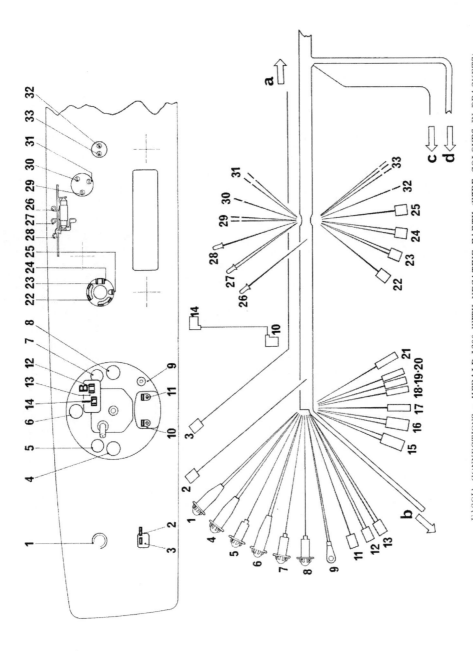

FASCIA CONNECTIONS - HERALD 1200 WITH NEGATIVE EARTH (WIRE COLOURS IN BRACKETS)

1 (LG/P and B) Bulb holder—flasher warning light. 2 (W) Lucar—heater switch. 3 (NW) Lucar—heater. 4 (W and Y) Bulb holder—instrument—ignition warning light. 5 (RW) Bulb holder—instrument illumination. 6 (W and WN) Bulb holder—instrument oil pressure warning light. 7 (RW) Bulb holder—instrument illumination. 8 (VW) Bulb holder—instrument main beam warning light. 9 (B) Eyelet 2 wire instrument. 10 (G) Lucar—fuel indicator. 11 (GB) Lucar—fuel indicator. 12 (W) Lucar—2 wire—voltage stabiliser. 13 (W) Lucar—2 wire. 14 (G) Lucar—voltage stabiliser. 15 (NR and R) Double snap connector—2 wire—column light switch. 16 (VW) Double snap connector—2 wire—column light switch. 17 (R) Lucar—snap connector—column light switch. 18 (LG/N)—flasher switch. 19 (GR)—flasher switch. 20 (GW) Flasher switch. 21 (NB) Snap connector—horn push. 22 (NU) Lucar—ignition starter switch. 23 (W) Lucar—2 wire—ignition/starter switch. 24 (W) Lucar—2 wire—ignition/starter switch. 25 (WR) Lucar—ignition/starter switch. 26 (NY) Terminal end—fascia lamp. 27 (NB) terminal end—2 wire. 28 (BG) Terminal end—fascia lamp. 29 (NU) Screw terminal—2 wire—master light switch. 30 (NR) Screw—master light switch. 31 (RW) Screw terminal—master light switch. 32 (BG) Screw terminal—windscreen wiper switch. 33 (B) Screw terminal—3 wire—windscreen wiper switch. a (NW) —to heater motor. b (RW)—to windscreen wiper motor. c (NB) —to RH door switch. d (W, BG, B and B) —to windscreen wiper motor.

COLOUR CODE

N Brown. U Blue. R Red. P Purple. G Green. LG Light Green. W White. Y Yellow. S Slate. B Black.

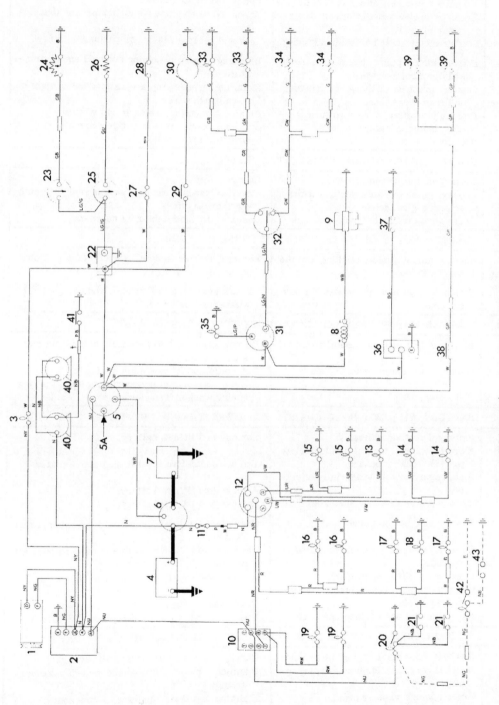

HERALD 13/60 NEGATIVE EARTH WIRING DIAGRAM

1 Generator. 2 Control box. 3 Ignition warning light. 4 Battery. 5 Ignition/starter switch. 5A Ignition/starter switch—radio supply connector. 6 Starter solenoid. 7 Starter solenoid. 8 Ignition coil. 9 Ignition distributor. 10 Master light switch. 11 Line fuse. 12 Column light switch. 13 Main beam warning light. 14 Main beam. 15 Dip beam. 16 Front parking lamp. 17 Tail lamp. 18 Plate illumination lamp. 19 Instrument illumination. 20 Fascia lamp. 21 Door switch. 22 Voltage stabiliser. 23 Fuel indicator. 24 Fuel tank unit. 25 Temperature indicator. 26 Temperature transmitter. 27 Oil pressure warning light. 28 Oil pressure switch. 29 Heater switch. 30 Heater motor. 31 Flasher unit. 32 Flasher switch. 33 LH flasher lamp. 34 RH flasher lamp. 35 Flasher warning light. 36 Windscreen wiper motor. 37 Windscreen wiper switch. 38 Stop lamp switch. 39 Stop lamp. 40 Horn. 41 Horn push. 42 Tailgate lamp switch (Estate only). 43 Tailgate lamp switch (Estate only).

COLOUR CODE

N Brown. U Blue. R Red. P Purple. G Green. LG Light Green. W White. Y Yellow. S Slate. B Black.

ELECTRICAL SYSTEM
FAULT FINDING CHART

Cause	Trouble	Remedy
SYMPTOM:	STARTER MOTOR FAILS TO TURN ENGINE	
No electricity at starter motor	Battery discharged	Charge battery.
	Battery defective internally	Fit new battery.
	Battery terminal leads loose or earth lead not securely attached to body	Check and tighten leads.
	Loose or broken connections in starter motor circuit	Check all connections and tighten any that are loose.
	Starter motor switch or solenoid faulty	Test and replace faulty components with new.
Electricity at starter motor: faulty motor	Starter motor pinion jammed in mesh with flywheel gear ring	Disengage pinion by turning squared end of armature shaft.
	Starter brushes badly worn, sticking, or brush wires loose	Examine brushes, replace as necessary, tighten down brush wires.
	Commutator dirty, worn, or burnt	Clean commutator, recut if badly burnt.
	Starter motor armature faulty	Overhaul starter motor, fit new armature.
	Field coils earthed	Overhaul starter motor.
SYMPTOM:	STARTER MOTOR TURNS ENGINE VERY SLOWLY	
Electrical defects	Battery in discharged condition	Charge battery.
	Starter brushes badly worn, sticking, or brush wires loose	Examine brushes, replace as necessary, tighten down brush wires.
	Loose wires in starter motor circuit	Check wiring and tighten as necessary.
SYMPTOM:	STARTER MOTOR OPERATES WITHOUT TURNING ENGINE	
Dirt or oil on drive gear	Starter motor pinion sticking on the screwed sleeve	Remove starter motor, clean starter motor drive.
Mechanical damage	Pinion or flywheel gear teeth broken or worn	Fit new gear ring to flywheel, and new pinion to starter motor drive.
SYMPTOMS:	STARTER MOTOR NOISY OR EXCESSIVELY ROUGH ENGAGEMENT	
Lack of attention or mechanical damage	Pinion or flywheel gear teeth broken or worn	Fit new gear teeth to flywheel, or new pinion to starter motor drive.
	Starter drive main spring broken	Dismantle and fit new main spring
	Starter motor retaining bolts loose	Tighten starter motor securing bolts. Fit new spring washer if necessary.
SYMPTOM:	BATTERY WILL NOT HOLD CHARGE FOR MORE THAN A FEW DAYS	
Wear or damage	Battery defective internally	Remove and fit new battery.
	Electrolyte level too low or electrolyte too weak due to leakage	Top up electrolyte level to just above plates
	Plate separators no longer fully effective	Remove and fit new battery.
	Battery plates severely sulphated	Remove and fit new battery.
Insufficient current flow to keep battery charged	Fan/dynamo belt slipping	Check belt for wear, replace if necessary, and tighten.
	Battery terminal connections loose or corroded	Check terminals for tightness, and remove all corrosion.
	Dynamo not charging properly	Remove and overhaul dynamo.
	Short in lighting circuit causing continual battery drain	Trace and rectify.
	Regulator unit not working correctly	Check setting, clean, and replace if defective.
SYMPTOM:	IGNITION LIGHT FAILS TO GO OUT, BATTERY RUNS FLAT IN A FEW DAYS	
Dynamo not charging	Fan belt loose and slipping, or broken	Check, replace, and tighten as necessary.
	Brushes worn, sticking, broken, or dirty	Examine, clean, or replace brushes as necessary.
	Brush springs weak or broken	Examine and test. Replace as necessary.
	Commutator dirty, greasy, worn, or burnt	Clean commutator and undercut segment separators.

ELECTRICAL SYSTEM

	Armature badly worn or armature shaft bent	Fit new or reconditioned armature.
	Commutator bars shorting	Undercut segment separations.
	Dynamo bearings badly worn	Overhaul dynamo, fit new bearings.
	Dynamo field coils burnt, open, or shorted.	Remove and fit rebuilt dynamo.
	Commutator no longer circular	Recut commutator and undercut segment separators.
	Pole pieces very loose	Strip and overhaul dynamo. Tighten pole pieces.
Regulator or cut-out fails to work correctly	Regulator incorrectly set	Adjust regulator correctly.
	Cut-out incorrectly set	Adjust cut-out correctly.
	Open circuit in wiring of cut-out and regulator unit	Remove, examine, and renew as necessary.

Failure of individual electrical equipment to function correctly is dealt with alphabetically, item by item, under the headings listed below:

FUEL GAUGE

Fuel gauge gives no reading	Fuel tank empty!	Fill fuel tank.
	Electric cable between tank sender unit and gauge earthed or loose	Check cable for earthing and joints for tightness.
	Fuel gauge case not earthed	Ensure case is well earthed.
	Fuel gauge supply cable interrupted	Check and replace cable if necessary.
	Fuel gauge unit broken	Replace fuel gauge.
Fuel gauge registers full all the time	Electric cable between tank unit and gauge broken or disconnected	Check over cable and repair as necessary.

HORN

Horn operates all the time	Horn push either earthed or stuck down	Disconnect battery earth. Check and rectify source of trouble.
	Horn cable to horn push earthed	Disconnect battery earth. Check and rectify source of trouble.
Horn fails to operate	Blown fuse	Check and renew if broken. Ascertain cause.
	Cable or cable connection loose, broken or disconnected	Check all connections for tightness and cables for breaks.
	Horn has an internal fault	Remove and overhaul horn.
Horn emits intermittent or unsatisfactory noise	Cable connections loose	Check and tighten all connections.
	Horn incorrectly adjusted	Adjust horn until best note obtained.

LIGHTS

Lights do not come on	If engine not running, battery discharged	Push-start car, charge battery.
	Light bulb filament burnt out or bulbs broken	Test bulbs in live bulb holder.
	Wire connections loose, disconnected or broken	Check all connections for tightness and wire cable for breaks.
	Light switch shorting or otherwise faulty	By-pass light switch to ascertain if fault is in switch and fit new switch as appropriate.
Lights come on but fade out	If engine not running battery discharged	Push-start car, and charge battery.
Lights give very poor illumination	Lamp glasses dirty	Clean glasses.
	Reflector tarnished or dirty	Fit new reflectors.
	Lamps badly out of adjustment	Adjust lamps correctly.
	Incorrect bulb with too low wattage fitted	Remove bulb and replace with correct grade
	Existing bulbs old and badly discoloured	Renew bulb units.
	Electrical wiring too thin not allowing full current to pass	Rewire lighting system.

167

ELECTRICAL SYSTEM

Cause	Trouble	Remedy
Lights work erratically - flashing on and off, especially over bumps	Battery terminals or earth connection loose Lights not earthing properly Contacts in light switch faulty	Tighten battery terminals and earth connection. Examine and rectify. By-pass light switch to ascertain if fault is in switch and fit new switch as appropriate.
WIPERS		
Wiper motor fails to work	Blown fuse Wire connections loose, disconnected, or broken Brushes badly worn Armature worn or faulty Field coils faulty	Check and replace fuse if necessary. Check wiper wiring. Tighten loose connections. Remove and fit new brushes. If electricity at wiper motor remove and overhaul and fit replacement armature. Purchase reconditioned wiper motor.
Wiper motor works very slowly and takes excessive current	Commutator dirty, greasy, or burnt Drive to wheelboxes too bent or unlubricated Wheelbox spindle binding or damaged Armature bearings dry or unaligned Armature badly worn or faulty	Clean commutator thoroughly. Examine drive and straighten out severe curvature. Lubricate. Remove, overhaul, or fit replacement. Replace with new bearings correctly aligned. Remove, overhaul, or fit replacement armature.
Wiper motor works slowly and takes little current	Brushes badly worn Commutator dirty, greasy, or burnt Armature badly worn or faulty	Remove and fit new brushes. Clean commutator thoroughly. Remove and overhaul armature or fit replacement.
Wiper motor works but wiper blades remain static	Driving cable rack disengaged or faulty Wheelbox gear and spindle damaged or worn Wiper motor gearbox parts badly worn	Examine and if faulty, replace. Examine and if faulty, replace. Overhaul or fit new gearbox.

CHAPTER ELEVEN

SUSPENSION – DAMPERS – STEERING

CONTENTS

General Description	1
Routine Maintenance	2
Inspecting the Suspension, Steering & Dampers for Wear	3
Anti-Roll Bar - Removal & Replacement	4
Coil Spring/Damper Units - Removal & Replacement	5
Rear Dampers - Removal & Replacement	6
Rear Semi-Elliptic Spring - Removal & Replacement	7
Rear Radius Arms - Removal & Replacement	8
Trunnion Housing Bushes - Removal & Replacement	9
Front Suspension Units - Removal & Replacement	10
Front Suspension - Dismantling & Reassembly	11
Front Wheel Alignment	12
Steering Wheel - Removal & Replacement	13
Steering Column - Removal & Replacement	14
Steering Column Bushes - Renewal	15
Rack & Pinion Steering Gear - Removal & Replacement	16
Rack & Pinion Steering Gear - Dismantling, Reassembly & Adjustment	17
Rack & Pinion Backlash - Adjustment	18
Outer Ball Joint - Removal & Replacement	19

SPECIFICATIONS

Front Suspension	Independent by coil springs & wishbones
Coil spring diameter	
Early 948 c.c. Heralds	3.543 to 3.587 in.
Later 948 c.c. Heralds	3.558 to 3.602 in.
Herald 1200, 12/50 & 13/60	3.13 + or - .02 in.
Free Length	
Early 948 c.c. Heralds	15.13 in.
Later 948 c.c. Heralds	11.7 in.
Herald 1200, 12/50 & 13/60	12.08 + or - .09 in.
Heavy duty & Courier van	10.97 in.
Length at 790 lbs load	
Early 948 c.c. Heralds	8.66 to 8.84 in.
Later 948 c.c. Heralds	7.79 to 7.89 in.
Herald 1200, 12/50 & 13/60	8.18 + or - .09 in.
Number of Working Coils	
Early 948 c.c. Heralds	$13\frac{1}{2}$
Later 948 c.c. Heralds	$9\frac{1}{2}$
Herald 1200, 12/50 & 13/60	$9\frac{1}{2}$
Camber angle	Nominal $2°$ positive) static laden
Castor angle	Nominal $4°$ positive) conditions
Steering axis inclination	$6\frac{3}{4}°$
Rear Suspension	Independent swing axle.
	Transverse leaf spring. Telescopic dampers

Number of spring leaves		Thickness of leaves
948 c.c. Herald Saloons	11	.2188 in.
948 c.c. Herald Coupes	8	.2188 in.

CHAPTER ELEVEN

Herald 1200, 12/50 & 13/60...	11	.2188 in.
Herald 1200 Coupe	8	.2188 in.
Estate cars	7	.31 in.
Courier van	8	.3125 in.
Heavy duty springs..	12	.2188 in.
Width of leaf springs	1.75 in.	

Spring test load   Spring rate

948 c.c. Herald saloon...	1420 lbs	270 lb/in.
948 c.c. Herald coupe	1010 lbs	202 lb/in.
Herald 1200, 12/50 & 13/60...	1420 lbs	270 lb/in.
Herald 1200 coupe...	1010 lbs	202 lb/in.
Estate cars	1735 lbs	510 lb/in.
Courier van	1910 lbs	552 lb/in.
Heavy duty springs..	1420 lbs	295 lb/in.
Steering...	Rack and pinion	
Steering wheel No. of turns lock to lock .	$3\tfrac{5}{8}$	
Pinion endfloat	.008 in.	
Damper endfloat	.004 in.	
Toe-in	0 to $\tfrac{1}{16}$ in. (0 to 1.6 mm.)	

Dampers
    Front & rear... ... ...   Telescopic double acting

Wheels & Tyres .. ..   Disc four stud fixing

Size	Wheels	Tyres
All Saloons, Convertibles & Coupe ...	$3\tfrac{1}{2}$ D	5.20 x 13 in.
Estate cars & Courier van	$4\tfrac{1}{2}$ J	5.60 x 13 in.

Pressures

		Front	Rear
All Saloons, Convertibles & Coupes	with 2 up	21 lb/sq.in. (1.48 kg/cm^2)	24 lb/sq.in. (1.7 kg/cm^2)
	with 3 up	21 lb/sq.in. (1.48 kg/cm^2)	28 lb/sq.in. (1.97 kg/cm^2)
Estate cars	with 2 up	21 lb/sq.in. (1.48 kg/cm^2)	25 lb/sq.in. (1.75 kg/cm^2)
	with 4 up	21 lb/sq.in. (1.48 kg/cm^2)	30 lb/sq.in. (2.1 kg/cm^2)
Courier van	Semi-laden ...	15 lb/sq.in. (1.06 kg/cm^2)	25 lb/sq.in. (1.76 kg/cm^2)
(4 ply tyres)	Fully-laden ...	15 lb/sq.in. (1.06 kg/cm^2)	32 lb/sq.in. (2.25 kg/cm^2)
(6 ply tyres)	Fully-laden ...	15 lb/sq.in. (1.06 kg/cm^2)	36 lb/sq.in. (2.53 kg/cm^2)
Torque wrench setting for wheel nuts...		38 to 42 lb/ft. (5.254 to 5.807 kg/cm)	

TORQUE WRENCH SETTINGS
REAR SUSPENSION

Rear damper lower attachment	30 to 32 lb/ft. (4.148 to 4.424 kg.m.)
Rear damper upper attachment	42 to 46 lb/ft. (5.807 to 6.360 kg.m.)
Spring to axle unit...	28 to 30 lb/ft. (3.871 to 4.148 kg.m.)
Spring ends to vertical link plate... ...	42 to 46 lb/ft. (5.807 to 6.360 kg.m.)
Vertical link plates to inner rear hub...	42 to 46 lb/ft. (5.807 to 6.360 kg.m.)

FRONT SUSPENSION

Anti-roll bar link assembly...	38 to 42 lb/ft. (5.254 to 5.807 kg.m.)
Anti-roll bar studs..	12 to 14 lb/ft. (1.659 to 1.936 kg.m.)
Anti-roll bar to chassis..	3 to 4 lb/ft. (0.281 to 0.415 kg.m.)
Ball assembly to upper wishbone... ...	16 to 18 lb/ft. (2.212 to 2.489 kg.m.)
Ball assembly to vertical link	38 to 42 lb/ft. (5.254 to 5.807 kg.m.)
Front damper - bottom...	42 to 46 lb/ft. (5.807 to 6.360 kg.m.)
Stub axle to vertical link	55 to 60 lb/ft. (7.604 to 8.295 kg.m.)
Top wishbone attachment	26 to 28 lb/ft. (3.595 to 3.871 kg.m.)
Trunnion to wishbone	35 to 38 lb/ft. (4.839 to 5.254 kg.m.)
Wishbone assembly to frame..	22 to 24 lb/ft. (3.042 to 3.318 kg.m.)
Vertical link to tie-rod lever..	32 to 35 lb/ft. (4.424 to 4.839 kg.m.)

SUSPENSION DAMPERS AND STEERING

STEERING GEAR

Coupling pinch bolts	18 to 20 lb/ft. (2.489 to 2.765 kg. m.)
Lower to upper clamp	6 to 8 lb/ft. (0.830 to 1.106 kg. m.)
Steering column safety clamp	6 to 8 lb/ft. (0.830 to 1.106 kg. m.)
Steering unit to frame	14 to 16 lb/ft. (1.936 to 2.212 kg. m.)

1. GENERAL DESCRIPTION

1. All four wheels are suspended independently on all models. At the front the suspension consists of two pairs of wishbones, the inner end of each pair being attached to and hinging on the chassis side members. The upper and lower suspension wishbones are fitted with nylon and rubber bushes and no greasing is therefore necessary.

2. Each stub axle carries the hub and road wheel and is attached by a nut and washer to the vertical link. Vertical links join each pair of wishbones at their outer ends. At the top of each link a ball joint and casing is held by two nuts and bolts between the front and rear arms of the wishbone. At the bottom of each link a bronze trunnion is held by a bolt, washer and nyloc nut between the outer ends of the lower wishbone.

3. A combined coil spring and telescopic damper unit is fitted between the outer end of the lower wishbone assembly and the bracket on the chassis frame which also supports the inner ends of the top wishbone. An anti-roll bar is fitted between the lower wishbones on each side.

4. The rack and pinion steering gear is held in place behind the radiator and above the chassis frame by inverted 'U' clamps at each end of the rack housing. Tie-rods from each end of the steering gear operate the steering arms via exposed and rubber gaiter enclosed ball joints. The upper splined end of the helically toothed pinion protrudes from the rack housing and engages with the splined end of the steering column. The pinion spline is grooved and the steering column is held to the pinion by a clamp bolt which partially rests in the pinion groove.

5. At the rear independent suspension is provided by a single transverse leaf spring dowelled and attached in its centre to the top of the differential unit housing to which it is clamped by a plate. At each end of the spring, two vertical links mounted on the spring eye carry the trunnion housing for the hub bearing and the swing axle shaft bearing assembly.

6. On the front face of the two vertical links a metalastik bushed radius arm is fitted to a bracket. The front end of the arm is fitted to a bracket on the chassis crossmember. In this way fore and aft movement of the rear suspension is firmly controlled. Telescopic rear dampers are fitted between the bottom of the vertical links and a bracket on the chassis frame.

2. ROUTINE MAINTENANCE

1. Suspension maintenance is much less than on most other cars. Every 6,000 miles remove the grease plug in the lower steering swivel and screw in a grease nipple. Fill a grease gun with Castrol Hypoy or a similar extreme pressure oil and pump in the lubricant until it wells out of the swivel. Unscrew the grease nipple and tighten down the plug. Then take the car to your local Triumph agent to have the toe-in checked.

2. At intervals of 12,000 miles undo the plug on the inside of the trunnion, replace it with a grease nipple, and give it exactly five strokes of the standard grease gun filled with Castrolease L.M. or similar. Overgreasing will result in the brake linings becoming contaminated, with a corresponding fall in braking efficiency.

3. The rack and pinion steering gear must also be greased at 12,000 mile intervals. Undo the small plug in the centre of the cap nut of the damper unit (immediately in front of the steering column where it enters the rack housing), fit a grease nipple, and give five strokes of the standard grease gun filled with Castrolease L.M. or similar. Over lubrication can make the rubber gaiters on the ends of the housing swell and split. On completion remove the nipple and refit the grease plug.

3. INSPECTING THE SUSPENSION, STEERING & DAMPERS FOR WEAR

1. To check for wear in the outer ball joints of the tie-rods place the car over a pit, or lie on the ground looking at the ball joints, and get a friend to rock the steering wheel from side to side. Wear is present if there is play in the joints.

2. To check for wear in the nylon and rubber bushes jack up the front of the car until the wheels are clear of the ground. Hold each wheel in turn, at the top and bottom and try to rock it. If the wheel rocks continue the movement at the same time inspecting the suspension to determine where play exists.

3. If the movement occurs between the wheel and the brake backplate then providing the wheel is on tightly the hub bearings will require replacement. The ball pin between the top suspension arms may be worn and movement here will be clearly seen, as will movement at the outer end of the lower wishbones.

4. How well the dampers function can be checked by bouncing the car at each corner. After each bounce the car should return to its normal ride position within $\frac{1}{2}$ to $\frac{3}{4}$ up-and-down movements. If the car continues to move up-and-down in decreasing amounts it means that the dampers are worn and must be replaced.

5. Excessive play in the steering gear will lead to wheel wobble, and can be confirmed by checking if there is any lost movement between the end of the steering column and the rack. Rack and pinion steering is normally very accurate and lost motion in the steering gear indicates a considerable mileage or lack of lubrication.

6. If backlash develops it is quite possible to take up the wear which will have occured between the teeth on the rack and the pinion by adjusting the pinion damper. (See Section 18).

7. The outer ball joints at either end of the tie-rods

CHAPTER ELEVEN

are the most likely items to wear first, followed by the rack ball joints at the inner end of the tie-rods.

8. At the rear end, bangs, clonks and squeaks can arise from a variety of sources and to determine the exact point of trouble is not difficult as long as a methodical and thorough check is made.

9. Start by checking the radius arm bushes for play, as they are of rubber they are fairly easy to replace. The damper bushes may also have worn and should be checked by heavy bouncing. Wear in the spring eye bush is unusual but can happen after high mileages.

10. Perhaps the commonest source of trouble are worn bushes in the trunnion housing. If with the trunnion housing bolt firmly tightened the trunnion housing can be seen to be moving loosely between the vertical links new bushes will have to be fitted. This is described in Section 9.

4. ANTI-ROLL BAR - REMOVAL & REPLACEMENT

1. Jack up the front of the car and place stands under the chassis for safety. Referring to Fig. 11.1. undo the nyloc nuts and washers (3,4) from the ends of the links (2).

2. The anti-roll bar is held to the chassis frame at two points by two 'U' bolts and clamps. Undo the nuts and washers (6,8) from the ends of the 'U' bolts (7) and pull out the clamps (10). Remove the anti-roll bar from the car.

3. Replacement is a straightforward reversal of the removal sequence. Do not tighten down the nuts fully until the car is on the ground and the suspension is in its normal laden position.

5. COIL SPRINGS, DAMPER UNITS - REMOVAL & REPLACEMENT

1. Loosen the nuts on the road wheel, jack up the front of the car and support it on stands, and take off the wheel. Place the wheel nuts in the hub cap for safe keeping. All numbers in brackets refer to Fig. 11.2.

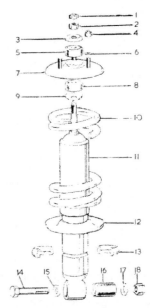

Fig. 11.2. EXPLODED VIEW OF A COIL SPRING/DAMPER UNIT
1 Locknut. 2 Nut. 3 Washer. 4 Nyloc nut. 5 Rubber bush. 6 Washer. 7 Upper spring pan. 8 Rubber bush. 9 Washer. 10 Road spring. 11 Damper. 12 Lower spring pan. 13 Collets. 14 Bolt. 15 Washer. 16 Metalastik bush. 17 Washer. 18 Nyloc nut.

2. Open the bonnet and then undo the nuts which hold the ends of the anti-roll bar to the lower wishbone. Undo and remove the three nuts and washers (4,6) which hold the upper spring pan (7) in place on the chassis sub frame.

3. From the lower end of the damper undo and remove the nut, bolt and washers (14,15,17,18) which holds the damper eye in place between the outer ends of the wishbone arms.

4. Slightly lift the brake drum and wishbones and pull out the coil spring/damper units from between the lower suspension arms.

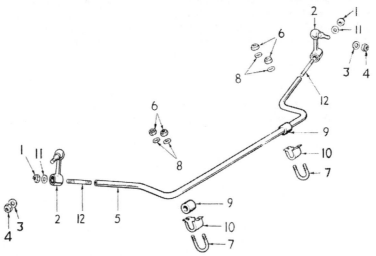

Fig. 11.1. EXPLODED VIEW OF THE ANTI-ROLL BAR
1 Nyloc nut. 2 Link. 3 Plain washer. 4 Nyloc nut. 5 Anti-roll bar. 6 Nyloc nut. 7 'U' bolt. 8 Plain washer. 9 Rubber bush. 10 Clamp. 11 Plain washer. 12 Stud.

SUSPENSION – DAMPERS – STEERING

5. Removing the coil spring from the damper unit involves the use of a heavy press and is definitely a job that should be left to a good garage. Under no circumstances try to remove the spring by using make shift equipment, as serious injury may result. Replacement is a straightforward reversal of the removal sequence. Always renew the rubber/metal bush (16) at the base of the damper unless the damper is virtually new.

6. **REAR DAMPERS - REMOVAL & REPLACEMENT**
1. Remove the hub cap, loosen the wheel nuts and jack up the rear of the car. Undo the nuts and take off the road wheel. Support the rear of the chassis on suitable stands.
2. Place the jack under the crossmember joining the two halves of the vertical links, and raise the jack so the load is taken off the damper.
3. Undo and remove the nut and bolt from the upper eye of the damper where it is attached to the chassis. Then remove the nut and washer from the vertical link crossmember extension and pull the damper off.
4. Replacement is a straightforward reversal of the removal sequence. Always fit new rubber bushes to the top eye of the damper and bleed the damper before fitment by operating it vertically over its full travel. Hold the damper vertically while fitting it.

7. **REAR SEMI-ELLIPTIC SPRING - REMOVAL & REPLACEMENT**
1. Loosen the rear wheel securing nuts, jack up the rear of the car and place support blocks or stands under the chassis frame. Remove the road wheels. All numbers refer to Fig. 11.5.
2. Disconnect the flexible hydraulic pipes from the chassis bracket as described in Chapter 9/6, and disconnect the handbrake cable by removing the clevis pin from the lever on the backplate. Remove the small pull-off spring from the backplate.
3. Remove the rear dampers as described in Section 6.
4. Mark a mating line across the drive shaft universal joint flanges (45) and undo the four nuts and bolts (43, 47) which secure the universal joint to the differential drive shaft.
5. Undo the nut (11) and remove the bolt (46) from each road spring eye (A). (Photo).

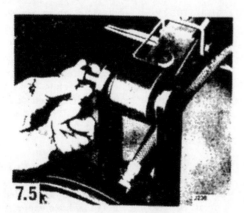

6. From behind the front seats inside the car remove the two screws which hold the cover in place over the centre of the spring and then undo the six nuts and washers (4, 5) from the studs which hold the spring in place (photo).

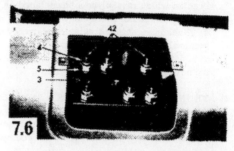

7. Lift off the spring clamp plate (3) and undo the three rear studs (42) from the differential casing.
8. The road spring can now be pulled away from the side of the car (photo).

9. In general replacement is a straightforward reversal of the removal sequence but note the following points.
a) The spring must be fitted with the edge marked 'FRONT' towards the front of the car.
b) The studs are fitted with their shorter threaded ends into the differential casing.
c) To prevent any water entering the car through the spring access cover smear the edge with 'PRESTIK', and with the cover in place smear the joint liberally with Seelastik.
d) Do not tighten fully the spring eye and damper mounting nuts and bolts until the car is on the ground in its static laden condition. This is to allow the rubber bushes to assume their normal working positions.

8. **REAR RADIUS ARMS - REMOVAL & REPLACEMENT**
1. Loosen the road wheel nuts, jack up the rear of the car, place stands under the rear of the chassis and remove the road wheels. All numbers in brackets refer to Fig. 11.5.
2. Place a jack under the crossmember at the base of the vertical links (10) and adjust the jack until the radius arm attachment bolts (33, 50) can be pulled out.

173

3. If the old rubber bushes are worn, cut, or perished, push them out and press in new ones by sandwiching them in a vice. If the holes in the bracket (34) are worn it will be necessary to fit a new bracket. Note the shims (35) fitted under the bracket which must always be replaced on reassembly to give the correct rear wheel toe-in of $\frac{1}{16}$ in.

4. Replacement is a straightforward reversal of the removal sequence. It is best to tighten the radius arm nuts and bolts down finally, when the car is in its normal static laden condition.

9. TRUNNION HOUSING BUSHES - REMOVAL & REPLACEMENT

1. It is not necessary to remove the trunnion housing from the car (26) to fit new steel and nylon bushes (29, 27).

2. Place a jack under the vertical links (10) and take out the bolt (50) which holds the radius arm (32) to its bracket.

3. Undo the nut and washer (13, 14) which holds the bottom of the damper to the vertical link crossbar extension, loosen the top damper fixing bolt (44) and pull the bottom of the damper (9) away from the extension.

4. Access can now be gained to the trunnion bolt (31) which is unscrewed from its nut and then pulled out of the trunnion housing. Lower the jack and push the bottom of the vertical links in towards the differential to expose the sides of the trunnion housing.

5. Knock out the steel bush (29) and prise out the nylon bushes (27). This can be a very difficult job if the bushes are worn and badly deformed.

6. Generously grease the new nylon bushes (27) with Retinax 'A' or a similar lithium based grease and fit one bush into place either side of the trunnion housing. Tap the bushes fully home with a hide faced hammer if need be.

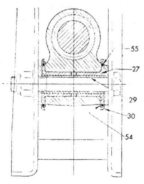

Fig. 11.3. A SECTIONED VIEW OF THE TRUNNION BUSHES
27 Nylon bush. 29 Steel bush. 30 Dust seal cover. 54 Dust seal. 55 Rubber ring.

7. Then fit the rubber grease sealing ring (55) in the groove between the flange of the nylon bush and the housing, followed by the grease seal cover (30).

8. Place the jack under the vertical link crossmember and jack the links up against the pressure of the spring until the hole for the trunnion housing bolt lines up with the holes in the vertical links. Insert the bolt and tighten the nut.

9. Replacement is now a straightforward reversal of the removal sequence.

10. FRONT SUSPENSION UNITS - REMOVAL & REPLACEMENT

1. Each front suspension unit comprising the top and bottom wishbones, coil spring/damper unit, vertical link, hub, brake assembly and the bracket on which the inner ends of the wishbones pivot can be removed from the car as a complete assembly.

2. Loosen the road wheel nuts, jack up the front of the car and place jacks under the chassis frame, and take off the road wheels.

3. If taking off the front suspension unit from the drivers side loosen the impact clamp bolts on the steering column and pull the steering column from the coupling.

4. Screw the brake reservoir cap down hard over a thin piece of polythene sheeting to prevent the hydraulic fluid draining away and disconnect the hydraulic flexible hoses from the brackets or side valance. (See Chapter 9, Section 6).

5. On models fitted with side valances undo the nut and bolt which holds the valance in place against the side frame.

6. Free the link on the end of the anti-roll bar from its attachment on the lower wishbone.

7. Undo the nut and washer from the steering arm ball joint and free the ball joint shank from the steering arm using an extractor or by impact hammering.

8. Referring to Fig. 11.8. undo the nyloc nuts and washers (38, 39) which hold the fulcrum brackets (34, 35) to the chassis frame. Note carefully the number and position of the shims (40) behind the brackets.

9. Referring to Fig. 11.4. undo the four bolts (1), spring washers and plates, from the outer face of the subframe and the single bolt (2) which holds the inner end of the subframe to the chassis frame. The complete suspension unit on one side of the car is now free to be lifted away from the chassis.

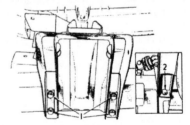

Fig. 11.4. THE FRONT SUSPENSION SUBFRAME ATTACHMENT POINTS
1 Outer bolts. 2 Inner bolts.

The other side is removed in identical fashion.

10. Reassembly is a straightforward reversal of the removal sequence. Make certain that the shims occupy their original positions, and on completion take the car to your local TRIUMPH dealer to have the castor and camber angles, and the front wheel alignment checked.

11. FRONT SUSPENSION - DISMANTLING & REASSEMBLY

1. The front suspension can be dismantled either on or off the chassis. In the authors experience it will be found easier to dismantle the suspension with the suspension unit still attached to the chassis frame.

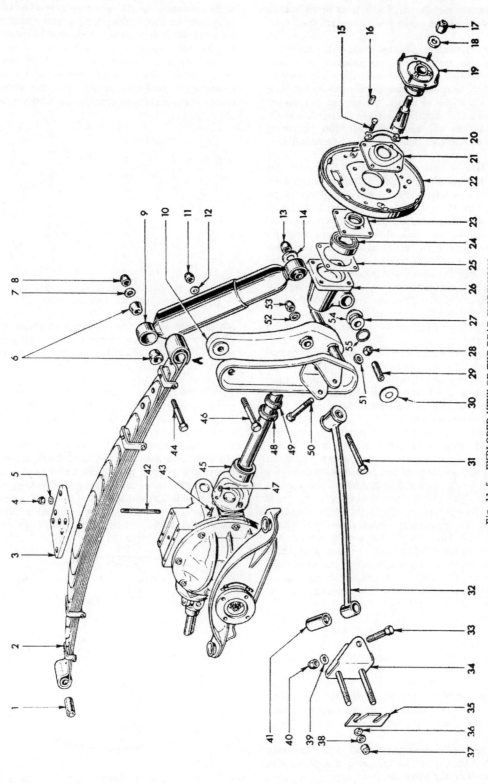

Fig. 11.5. EXPLODED VIEW OF THE REAR SUSPENSION

1 Spring eye bush. 2 Road spring. 3 Spring clamp plate. 4 Nut. 5 Washer. 6 Rubber bush. 7 Washer. 8 Nut. 9 Damper. 10 Vertical link. 11 Nut. 12 Washer. 13 Nut. 14 Washer. 15 Bolt. 16 Key. 17 Nut. 18 Washer. 19 Hub. 20 Lock tab. 21 Grease retainer. 22 Brake backplate. 23 Seal housing. 24 Bearing. 25 Gasket. 26 Trunnion housing. 27 Nylon bush. 28 Nut. 29 Steel bush. 30 Dust seal cover. 31 Bolt. 32 Radius arm. 33 Bolt. 34 Radius arm bracket. 35 Shim. 36 Washer. 37 Nut. 38 Washer. 39 Washer. 40 Nut. 41 Rubber bush. 42 Stud. 43 Bolt. 44 Bolt. 45 Drive shaft coupling. 46 Bolt. 47 Nut. 48 Flinger. 49 Seal. 50 Bolt. 51 Washer. 52 Washer. 53 Nut. 54 Dust seal. 55 Rubber ring.

175

CHAPTER ELEVEN

2. Follow the instructions as far as paragraph 4 in Section 5 (coil spring/damper unit - removal & replacement). Remove the anti-roll bar as described in Section 4.

3. On Herald models fitted with front drum brakes undo the retaining screws and take off the brake drums.

Take off the brake calliper assembly after undoing the two bolts and spring washers nearest to the vertical link (10) which hold the calliper in place. If the suspension has not been removed from the chassis frame as a unit, tie the calliper up out of the way so no strain is placed on the flexible hose.

4. Pull off the hub assembly (20) after removing the grease cap (26), split pin (25), and castellated nut (24).

5. Undo and remove the four nuts and bolts (28, 29) and the nyloc nut and washers (15, 16) which hold the calliper mounting bracket and dust shield to the vertical link (10) and the steering arm (27). Push the stub axle (14) out of the vertical link.

6. To free the inner ends of the upper wishbones (1) from the pivot point on the chassis undo the two bolts (3) and nyloc nuts (4) which hold them in place.

7. Then free the lower wishbone brackets (34, 35) from the chassis frame by undoing the retaining nuts and washers (38, 39). Note the number and the position of the shims (40). The wishbone arms together with the vertical link can now be removed from the car as a unit.

8. To separate the upper wishbone ball joint assembly (5) from the vertical link (10) undo the nyloc nut (12) not more than 1½ turns and hit the head of the nut sharply with a hide hammer to free the taper on the shank of the ball joint from the taper in the link. Impact hammering can also help at this stage. If the ball joint refuses to move then take off the nyloc nut and borrow an extractor to separate the ball joint and link. The ball joint is sandwiched between the outer ends of the front and rear upper suspension arms (1), and is simply detached by undoing the two retaining bolts, nuts and washers (7, 8, 9).

9. Next free the outer ends of the lower wishbone (30) from the trunnion (41) on the bottom of the vertical link (10). To do this simply remove the nut and washer (48, 49) from the bolt (47) which holds the trunnion in place between the wishbones and pull out the bolt. The steel bush (44), nylon bushes (42) and associated components can then be pushed out of the trunnion.

10. To separate the trunnion (41) from the vertical link (10) unscrew the former and then pull off the grease seal (50).

11. If the suspension is being fully dismantled and the car has covered in excess of 30,000 miles it is well worthwhile renewing all the rubber bushes (2, 53, 31) as a matter of course. Press the old bushes out in a vice and press in the new onces so their ends protrude equally either side of the wishbone eyes.

12. Reassembly commences by fitting the stub axle (14) to the vertical link and securing it in place with a plain washer and nyloc nut.

13. Rebuild the trunnion using new nylon bushes (42) and remember to fit the steel sleeve (41) and then the flanged washer (43) with their flanges facing away from the trunnion. When reassembled peel the dust excluder (45) over the bushes and trunnion flange to keep out road dirt.

14. Fit a new oil seal (50) to the thread at the bottom of the vertical link (10) and then screw the trunnion fully onto the link. From this position unscrew it to the first working position. It is most important that the left-hand threaded link and trunnion are fitted on the left-hand side of the car and similarly that the right-hand threaded link and trunnion are fitted to the right-hand side of the car. The right-hand trunnion has a reduced diameter at its lower end as shown in Fig. 11.6.

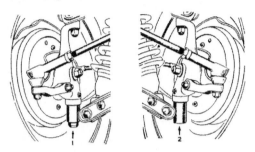

Fig. 11.6. 1. The right-hand lower trunnion (with reduced diameter lower end). 2. Left-hand trunnion.

15. Reassemble the trunnion between the outer ends of the lower wishbone, remembering to fit the larger flanged washers (46) with their flanges towards the trunnion. Insert the bolt (47) and do up the securing nut (49).

16. Fit the fulcrum brackets (32) to the inner ends of the lower wishbones with the bracket with the longest portion above the fixing stud fitted to the rear wishbone arm.

17. The ball joint assembly can now be fitted to the top wishbones and to the vertical link followed by the steering arm (27), and brake backplate or dust shield. The latter must be sealed to the calliper bracket with Expandite Seal-a-Strip as shown in Fig. 11.7. On some models a rubber seal will be found fitted between the recessed face of the calliper brakcet and the vertical link.

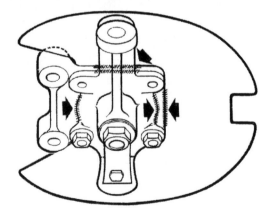

Fig. 11.7. Seal the disc brake calliper bracket to the vertical link with Expandite Seal-a-strip in the shaded areas.

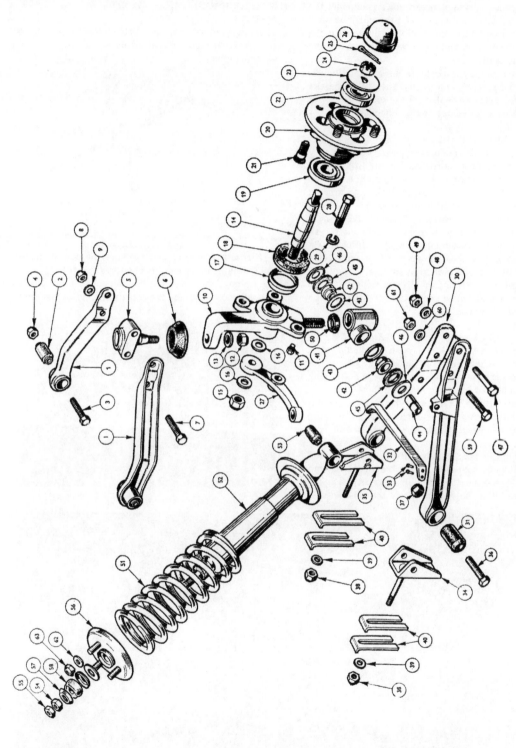

Fig. 11.8. EXPLODED VIEW OF THE FRONT SUSPENSION

1 Top wishbone arm and bush assembly. 2 Rubber bush. 3 Bolt. 4 Nyloc nut. 5 Upper wishbone ball joint assembly. 6 Gaiter. 7 Bolt. 8 Nyloc nut. 9 Plain washer. 10 Vertical link. 11 Grease plug. 12 Nyloc nut. 13 Plain washer. 14 Stub axle. 15 Nyloc nut. 16 Plain washer. 17 Water shield. 18 Oil seal. 19 Inner hub bearing. 20 Front hub and stub assembly. 21 Wheel stud. 22 Outer hub bearing. 23 'D' washer. 24 Slotted nut. 25 Split pin. 26 End hub cover. 27 Tie rod lever. 28 Bolt—tie rod lever to vertical link. 29 Spring washer. 30 Lower wishbone assembly. 31 Rubber bush. 32 Lower wishbone strut. 33 Rivet—strut to wishbone. 34 Lower wishbone fulcrum bracket—front. 35 Lower wishbone fulcrum bracket—rear. 36 Bolt—fulcrum brackets to wishbone. 37 Nyloc nut. 38 Nyloc nut—fulcrum bracket to frame. 39 Plain washer. 40 Shim. 41 Bottom trunnion assembly. 42 Bearing. 43 Road spring. 44 Distance piece. 45 Lower trunnion dirt seal. 46 Water shield. 47 Bolt—trunnion to wishbone. 48 Plain washer. 49 Nyloc nut. 50 Oil seal. 51 Road spring. 52 Shock absorber. 53 Rubber bush. 54 Nut. 55 Lock nut. 56 Upper spring assembly plate. 57 Cap. 58 Rubber bush. 59 Bolt—lower shock absorber attachment. 60 Plain washer. 61 Nyloc nut. 62 Plain washer. 63 Nyloc nut—shock absorber upper attachment.

18. Reassemble and adjust the hub as described on page 140., and then fit the wishbones to the chassis and subframe remembering to fit the shims behind the lower brackets. Do not fully tighten the nuts on the bolts 3, 36 and 59 yet.

19. Replace the spring/damper unit, reconnect the tie-rod to the steering arm, refit the calliper assembly, reconnect the anti-roll bar, lubricate the vertical link trunnion and then lower the front of the car to the ground.

20. With the front of the car at its normal ride height tighten the nuts, 4, 37 and 61 and then take the car to your local Triumph dealer for the toe-in to be set and the castor and camber angles checked.

12. FRONT WHEEL ALIGNMENT

1. The front wheels are correctly aligned when they turn in at the front not more than 1/16 in. Adjustment is effected by loosening the locknut on each tie-rod ball joint, and the clips on the gaiters, and turning both tie-rods equally until the adjustment is correct.

2. This is a job that your local Triumph agent must do, as accurate alignment requires the use of expensive base bar or optical alignment equipment. On no account try to do this job yourself, using planks of wood or other make shift implements.

3. If the wheels are not in alignment, tyre wear will be heavy and uneven, and the steering will be stiff and unresponsive.

13. STEERING WHEEL - REMOVAL & REPLACEMENT

1. Carefully prise off the horn push button (63). (All numbers in brackets refer to Fig.11.10) (photo).

2. Lift the horn button away to expose the steering wheel securing nut (photo).

3. Next pull out the horn contact brush (61), (photo)

4. Undo and remove the steering wheel securing nut (62) and the special clip (60) which lies underneath it (photo).

5. The wheel is then simply pulled off the splines (photo). Replacement is a simple reversal of this process. To ensure the wheel is correctly aligned on refitment it is helpful to mark the boss and a corresponding spline before pulling the wheel off. Tighten the securing nut down firmly and peen the metal of the nut to the inner column to prevent the nut working loose.

14. STEERING COLUMN - REMOVAL & REPLACEMENT

1. Undo and remove the bolt at the bottom of the clamp on the column immediately adjacent to the rack housing

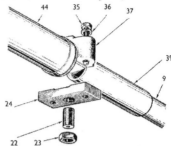

Fig.11.9. THE COMPONENT PARTS OF THE STEERING COLUMN IMPACT CLAMP
9 Lower column. 22 Socket screw. 23 Nut. 24 Clamp plate. 35 Bolt. 36 Spring washer. 37 Clamp. 39 Inner upper column. 44 Outer upper column.

2. Separate the cables on the steering column under the fascia at their snap connectors (photo).

3. Remove the outer column support clamp (photo) halfway down the column after undoing the securing bolts.

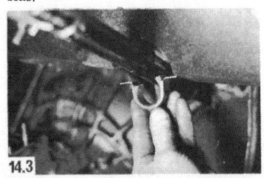

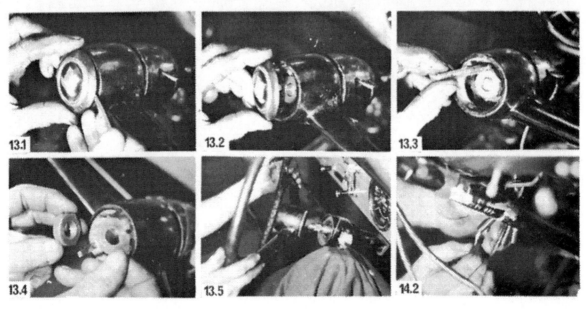

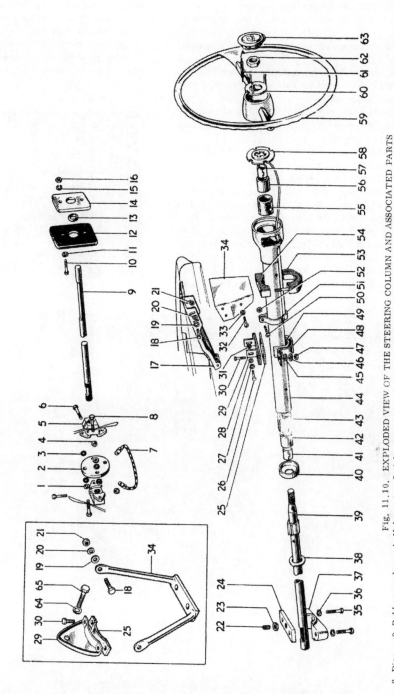

Fig. 11.10. EXPLODED VIEW OF THE STEERING COLUMN AND ASSOCIATED PARTS

1 Washer. 2 Disc. 3 Rubber washer. 4 Nyloc nut. 5 Adaptor. 6 Pinch bolt. 7 Earth cable. 8 Bolt. 9 Lower steering column. 10 Bolt. 11 Washer. 12 Rubber seal. 13 Washer. 14 Retaining plate. 15 Spring washer. 16 Nut. 17 Support bracket. 18 Bolt. 19 Spring washer. 20 Washer. 21 Nut. 22 Socket screw. 23 Nut. 24 Clamp plate. 25 Felt pad. 26 Bolt. 27 Spring washer. 28 Washer. 29 Bracket. 30 Bolt. 31 Nut. 32 Screw. 33 Washer. 34 Bracket. 35 Bolt. 36 Spring washer. 37 Clamp. 38 Nylon washer. 39 Upper inner steering column. 40 End cap. 41 Nylon bush. 42 Steel bush. 43 Rubber bush. 44 Outer upper column. 45 Washer. 46 Spring washer. 47 Nut. 48 Lower outer column clamp. 49 Felt pad. 50 Screw. 51 Cable trough clip. 52 Nut. 53 Upper clamp (lower half). 54 Upper clamp (upper half). 55 Rubber bush. 56 Steel bush. 57 Nylon bush. 58 Horn contact brush. 59 Steering wheel. 60 Clip. 61 Horn contact ring. 62 Nut. 63 Horn push. 64 Spring washer. 65 Bolt.

Inset shows upper outer column clamp attachment on Herald 1200, 12/50 and Vitesse.

CHAPTER ELEVEN

4. Then undo the two bolts which hold the upper clamp in place and remove the clamp from the column.

5. Remove the special clip which holds the cables to the column.

6. To avoid damage to the switches on the sides of the column detach them after removing the switch covers by simply undoing the two small screws which hold each switch in place (photo).

7. Carefully lift the switches away from the column (photo).

8. Undo and remove the steering wheel as described in Section 13.

9. At the base of the column inside the car, remove the two bolts which hold the two halves of the impact clamp together (photo).

10. The lower column can then be removed from the engine compartment and the upper column from inside the car (photo).

11. Replacement is a straightforward reversal of the removal sequence but note the following points.

When fitting the steering wheel the two lips on the direction indicator cancelling lug on the column must be parallel and in line with the horizontal spokes of the wheel. With the wheel at the desired height tighten the bolts (35) on the impact clamp (37). Note that the column will not be able to telescope if adjusted to its lowest position. Tighten the allen screw with an Allen key by hand as far as possible without bending the latter. Then tighten the locknut (23).

15. STEERING COLUMN BUSHES - REMOVAL

1. If there is any play in the top of the steering column it will be necessary to replace the rubber and nylon bush at the top and possibly also the similar bush at the bottom of the outer column. Pull the inner from the outer column (photo).

2. At the lower end of the column take off the end cap, depress the buttons on the rubber bushes where they protrude through the hole in the bottom and the top of the column and with a steel bar carefully drift the bushes out (photo).

3. As soon as the bushes are removed from the top and bottom of the column (photo) the metal and nylon inserts can be taken from them, unless completely new units (strongly advised) are fitted on reassembly.

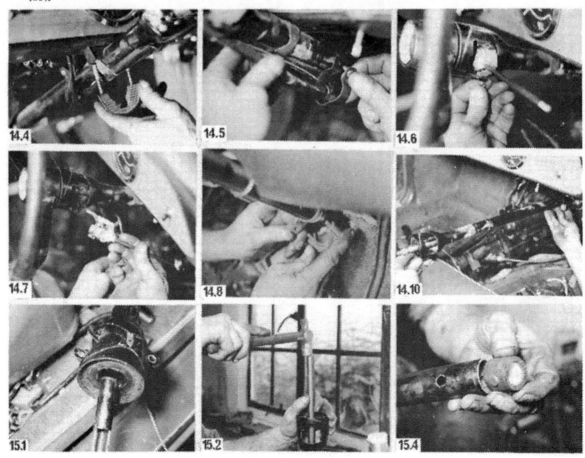

SUSPENSION – DAMPERS – STEERING

4. If the old steel sleeves and nylon bushes are being used fit them to new rubber bushes, or fit a completely new bush assembly (photo) to the column, making sure the metal reinforcement ring at the end of the bush is located towards the lower end of the column and that the protrusion on the rubber bushes engage in the locating holes in the column.

5. Fit the end cap to the lower end of the column.

16. RACK & PINION STEERING GEAR - REMOVAL & REPLACEMENT

1. All numbers in brackets refer to Fig. 11.12. Unscrew the nut (8) and remove the bolt (9) from the bottom of the clamp (7) on the lower end of the steering column. Undo the bolt from the chassis which holds the earth lead running to the grease plug (28) on the rack and pinion steering unit (25).

2. Remove the nut (19) from each of the two tie-rod to steering arm ball joints (44) and unscrew the nuts two turns. Hit the head of the nut with a soft faced hammer to drive the shank of the ball joint out of the steering arm. Remove each nut and pull the tie-rods complete with ball joints away.

3. Undo the nuts and washers (24) from the inverted 'U' bolts (20), remove the locating plates (23) and the rubber bushes (21).

4. Pull the steering gear sufficiently forward to disengage the pinion shaft (15) from the clamp (7) and remove the steering gear from the side of the car.

5. Replacement commences by centralising the pinion shaft. Turn the shaft from lock to lock counting the number of turns and then rotate it half this number. Ensure the steering wheel is in the straight ahead position.

6. Offer the unit up to the chassis, connect the clamp to the splined pinion shaft, fit the rubber bushes, 'U' bolts, plates and nyloc nuts loosely and ensure the steering gear is positioned as shown in Fig. 11.11. Then slide the 'U' bolt assemblies outwards until a clearance of $\frac{1}{8}$ in. is present between the flanges on the 'U' bolt retainers and the rack and pinion flanges.

7. The 'U' bolts must then be held in this position while a friend slides the plates (23) inwards until their flanged faces abut the chassis. Then tighten the 'U' bolt securing nuts.

8. Reconnect the earth strap, fit the clamp nut and bolt, refit the tie-rods to the steering arms, and then take the car to your local garage or Triumph agent to have the toe-in checked.

17. RACK & PINION STEERING GEAR - DISMANTLING, REASSEMBLY & ADJUSTMENT

1. All numbers in brackets refer to Fig. 11.12. Undo the clips (40, 42) on the rubber gaiters (41) and pull the gaiters back to expose the inner ball joint assemblies.

2. Loosen the locknuts (33), and completely unscrew both the tie-rod assemblies (38) from the rack (32).

3. Pull out the coil springs (36), free the tab washer (35), unscrew the nut (34) and take off the tab washer, shims (26) and cup (37).

4. Mark the positions of the locknuts (43) on the tie-rods (38) so the 'toe-in' is approximately correct on reassembly, undo the locknuts (43) and screw off the ball joints.

5. Pull off the rubber gaiter (41) and undo the cup nut (39) from the tie-rods, and then remove the locknuts (33).

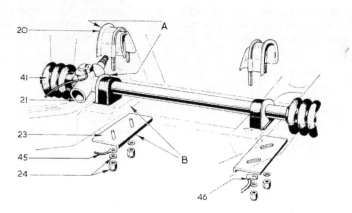

Fig. 11.11. Ensure the steering gear is fitted in the position indicated in the following measurements:—
A Distance between flanges must be $\frac{1}{8}$ in. (3.17 mm.)
B Flange of item (23) must contact innermost flange of frame
20 'U' bolt. 21 Rubber bush. 23 Locating plates. 24 Nyloc nuts. 41 Rubber gaiter. 45 Steering column earth cables. 46 Engine earth cables.

6. Unscrew and remove the damper cap (27), spring (30), plunger (31) and shims (29). Fig.11.12 refers).

7. The pinion can be removed after taking out the circlip (10). Take great care not to lose the small locating peg (5).

8. The rack can now be pulled from the housing tube and the thrust washer (16) and bush (17) taken from the pinion housing (22).

9. Clean all the parts thoroughly and examine the teeth of the rack and pinion for wear or damage. Also examine the pinion thrust washers, and bushes, thrust pad, and inner and outer ball joints and replace as necessary.

10. Reassembly commences by sliding the rack (32) into place. Then fit the bush (17) and thrust washer (16) and adjust the pinion endfloat after assembling the thrust washer (14), bush (13) and retaining ring (11) to the pinion, and fitting and securing the pinion to the housing with the circlip.

11. By trial and error fit a different number of shims (12) (available in thicknesses of 0.004 in. and 0.016 in.) until the minimum amount of endfloat exists when the pinion is pushed in and out commensurate with free rotation of the pinion.

12. When the correct shim thickness has been ascertained fit a new rubber 'O' ring to the retaining ring (11), and after the pinion assembly has been fitted, fit the dowel (5) and circlip (10).

13. Then adjust the rack and pinion backlash as described in Section 18. Reassembly is now a straightforward reversal of the dismantling sequence. Check that the rack ball joints fit tightly but are free to move. If they are excessively loose or tight adjustment can be made by varying the number and thickness of the shims (26) between the ball joint cup (37) and the ball housing.

14. After fitting the rack and pinion in place fill it with the recommended lubricant.

18. RACK & PINION BACKLASH - ADJUSTMENT

Backlash between the pinion and the rack can be taken up by means of an adjustment between the rack damper cap, and the rack housing. If backlash is present adjust the rack damper in the following manner. Numerical references in brackets refer to Fig. 11.12.

1. Disconnect the outer ends of the steering tie-rods (44) by knocking out the tie-rod ball joint shanks from the holes in the steering arms.

2. Unscrew the damper cap (27), and remove the spring (30) and shims (29).

3. Refit the damper cap together with the plunger (31), but without the spring and the shims.

4. Tighten the damper cap until it requires about a 2 lb pull at the steering wheel rim to turn the wheel.

5. Measure the gap between the underside of the damper cap and the rack housing with a feeler gauge, and add to this figure a clearance figure of .002 to .005 in. (.05 to .127 mm.). The total figure represents the thickness of the shims that must be fitted under the cap. Shims are available in thicknesses of .003 in. and .010 in. (.76 and .254 mm.).

6. Remove the cap, replace the spring, fit the necessary shims and tighten the cap down firmly. Reconnect the tie-rod ball joints to the steering arms and note the improvement when the car is taken on the road.

19. OUTER BALL JOINT — REMOVAL & REPLACEMENT

If the tie-rod outer ball joints are worn it will be necessary to renew the whole ball joint assembly as they cannot be dismantled and repaired. To remove a ball joint, free the ball joint shank from the steering arm and mark the position of the locknut on the tie-rod accurately to ensure near accurate 'toe-in' on reassembly.

Slacken off the ball joint locknut, and holding the tie-rod by its flat with a spanner, to prevent it from turning, unscrew the complete ball assembly from the rod. Replacement is a straightforward reversal of this process. Visit your local Triumph agent to ensure that toe-in is correct.

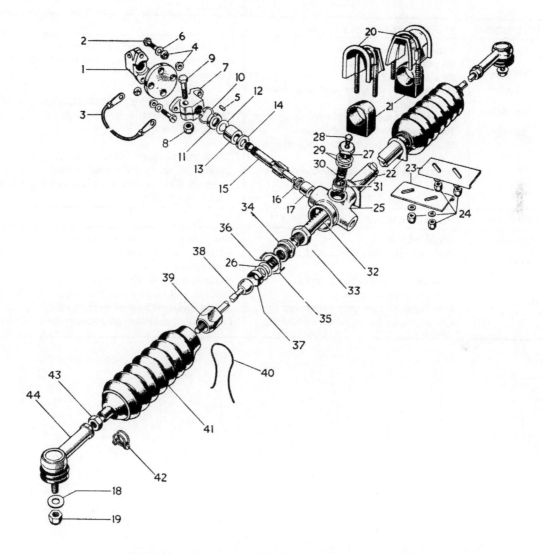

Fig. 11.12. EXPLODED VIEW OF THE RACK AND PINION STEERING
1 Upper steering coupling. 2 Bolt. 3 Earth cable. 4 Rubber bushes. 5 Dowel. 6 Plain washer. 7 Lower steering coupling. 8 Nyloc nut. 9 Pinch bolt. 10 Circlip. 11 Retaining ring. 12 Shims. 13 Upper pinion bush. 14 Thrust washer. 15 Pinion shaft. 16 Thrust washer. 17 Lower pinion bush. 18 Plain washer. 19 Nyloc nut. 20 'U' bolts—rack and pinion body to frame. 21 Rubber bushes. 22 Abutment plates. 23 Locating plates—rack and pinion body to frame. 24 Nyloc nuts. 25 Rack and pinion body assembly. 26 Shims. 27 Damper cap. 28 Grease plug. 29 Shims. 30 Damper spring. 31 Plunger. 32 Rack. 33 Locknut. 34 Sleeve nut. 35 Tab washer. 36 Spring. 37 Ball socket cup. 38 Tie rod. 39 Cup nut. 40 Locking wire. 41 Bellows gaiter. 42 Gaiter clip. 43 Locknut. 44 Tie rod end ball joint assembly.

SUSPENSION – DAMPERS – STEERING

FAULT FINDING CHART

Cause	Trouble	Remedy
SYMPTOM:	**STEERING FEELS VAGUE, CAR WANDERS AND FLOATS AT SPEED**	
General wear or damage	Tyre pressures uneven Dampers worn or require topping up Spring clips broken Steering gear ball joints badly worn Suspension geometry incorrect Steering mechanism free play excessive Front suspension and rear axle pick-up points out of alignment	Check pressures and adjust as necessary. Top up dampers, test, and replace if worn. Renew spring clips. Fit new ball joints. Check and rectify. Adjust or overhaul steering mechanism. Normally caused by poor repair work after a serious accident. Extensive rebuilding necessary.
SYMPTOM:	**STIFF & HEAVY STEERING**	
Lack of maintenance or accident damage	Tyre pressures too low No grease in king pins No oil in steering gear No grease in steering and suspension ball joints Front wheel toe-in incorrect Suspension geometry incorrect Steering gear incorrectly adjusted too tightly Steering column badly misaligned	Check pressures and inflate tyres. Clean king pin nipples and grease thoroughly. Top up steering gear. Clean nipples and grease thoroughly. Check and reset toe-in. Check and rectify. Check and readjust steering gear. Determine cause and rectify (Usually due to bad repair after severe accident damage and difficult to correct).
SYMPTOM:	**WHEEL WOBBLE & VIBRATION**	
General wear or damage	Wheel nuts loose Front wheels and tyres out of balance Steering ball joints badly worn Hub bearings badly worn Steering gear free play excessive Front springs loose, weak or broken	Check and tighten as necessary. Balance wheels and tyres and add weights as necessary. Replace steering gear ball joints. Remove and fit new hub bearings. Adjust and overhaul steering gear. Inspect and overhaul as necessary.

CHAPTER TWELVE

BODYWORK AND UNDERFRAME

CONTENTS

General Description...	1
Maintenance - Body & Chassis..	2
Maintenance - Upholstery & Carpets	3
Maintenance - Hoods & Tonneau Covers...	4
Minor Body Repairs...	5
Major Chassis & Body Repairs..	6
Maintenance - Hinges & Locks..	7
Door Locks - Removal & Replacement	8
Striker Plate - Removal, Replacement & Adjustment	9
Door Rattles - Tracing & Rectification	10
Windscreen Glass - Removal & Replacement...	11
Instruments & Switches - Removal & Replacement...	12
Window Regulator Mechanism - Removal & Replacement	13
External Door Handles - Removal & Replacement...	14
Bonnet Assembly - Removal, Replacement & Adjustment	15
Bumpers - Removal & Replacement..	16
Heater - Removal & Replacement	17
Heater Fan Motor - Removal & Replacement..	18

1. GENERAL DESCRIPTION

An all steel body is bolted to a separate box section chassis frame, which consists of two longitudinal members joined together at three points with outrigger arms to carry the body (see Fig. 12.1).

The body consists of a number of sub-assemblies which are fixed together and then bolted to the chassis. The sub-assemblies are not welded together and this makes the removal and replacement of damaged panels a comparatively easy operation.

The three main body sub-assemblies consist of the front section; centre section, and rear section. The bonnet and front wings comprise the front section; the engine bulkhead, floor and windscreen frame the centre section; and the boot floor and rear wings the rear section.

All saloon and convertible models have two doors and the Courier van and Estate cars have a tailgate hinged at its upper end.

The doors are hinged at the front with built in door 'keeps' and are push button operated from the outside. External door locks are fitted to all models except the Herald 'S' where only the driver's door can be locked from outside.

The centre section of the body is the most important and is bolted to the chassis at six points. Herald body panels are specially treated with a rust preventative and unless the protective paint layer and preservative is damaged, no rust should form. Early examples of the Herald are frequently found to be in very good condition.

2. MAINTENANCE - BODY & CHASSIS

1. The condition of your car's bodywork is of considerable importance as it is on this that the second hand value of the car will mainly depend. It is much more difficult to repair neglected bodywork than to renew mechanical assemblies. The hidden portions of the body, such as the wheel arches and the underframe and the engine compartment are equally important, though obviously not requiring such frequent attention as the immediately visible paintwork.

2. Once a year or every 12,000 miles, it is a sound scheme to visit your local main agent and have the underside of the body steam cleaned. This will take about $1\frac{1}{2}$ hours and cost about £4. All traces of dirt and oil will be removed and the underside can then be inspected carefully for rust, damaged hydraulic pipe, frayed electrical wiring and similar maladies. The car should be greased on completion of this job.

3. At the same time the engine compartment should be cleaned in the same manner. If steam cleaning facilities are not available then brush 'Gunk' or a similar cleanser over the whole engine and engine compartment with a stiff paint brush, working it well in where there is an accumulation of oil and dirt. Do not paint the ignition system and protect it with oily rags when the Gunk is washed off. As the Gunk is washed away it will take with it all traces of oil and dirt, leaving the engine looking clean and bright.

4. The wheel arches should be given particular attention as undersealing can easily come away here and stones and dirt thrown up from the road wheels

can soon cause the paint to chip and flake, and so allow rust to set in. If rust is found, clean down the bare metal with wet and dry paper, paint on an anti-corrosive coating such as Kurust, or if preferred, red lead, and renew the paintwork and undercoating.

5. The bodywork should be washed once a week or when dirty. Thoroughly wet the car to soften the dirt and then wash the car down with a soft sponge and plenty of clean water. If the surplus dirt is not washed off very gently, in time it will wear the paint down as surely as wet and dry paper. It is best to use a hose if this is available. Give the car a final wash-down and then dry with a soft chamois leather to prevent the formation of spots.

6. Spots of tar and grease thrown up from the road can be removed with a rag dampened with petrol.

7. Once every six months, or every three months if wished, give the bodywork and chromium trim a thoroughly good wax polish. If a chromium cleaner is used to remove rust on any of the car's plated parts remember that the cleaner also removes part of the chromium so use sparingly.

3. MAINTENANCE - UPHOLSTERY & CARPETS

1. Remove the carpets or mats and thoroughly vacuum clean the interior of the car every three months or more frequently if necessary.

2. Beat out the carpets and vacuum clean them if they are very dirty. If the upholstery is soiled apply an upholstery cleaner with a damp sponge and wipe off with a clean dry cloth.

4. MAINTENANCE - HOODS & TONNEAU COVERS

Under no circumstances try to clean hoods and tonneau covers with detergents, caustic soaps, or spirit cleaners. Plain soap and water is all that is required with a soft brush to clean dirt that may be ingrained. Wash the hood as frequently as the rest of the car.

5. MINOR BODY REPAIRS

1. At some time during your ownership of your car it is likely that it will be bumped or scraped in a mild way, causing some slight damage to the body.

2. Major damage must be repaired by your local Triumph agent, but there is no reason why you cannot successfully beat out, repair and respray minor damage yourself. The essential items which the owner should gather together to ensure a really professional job are:-

a) A plastic filler such as Holts 'Cataloy'.
b) Paint whose colour matches exactly that of the bodywork, either in a can for application by a spray gun, or in an aerosol can.
c) Fine cutting paste.
d) Medium and fine grade wet and dry paper.

3. Never use a metal hammer to knock out small dents as the blows tend to scratch and distort the metal. Knock out the dent with a mallet or rawhide hammer and press on the underside of the dented surface a metal dolly or smooth wooden block roughly contoured to the normal shape of the damaged area.

4. After the worst of the damaged area has been knocked out, rub down the dent and surrounding area with medium wet and dry paper and thoroughly clean away all traces of dirt.

5. The plastic filler comprises a paste and a hardener which must be thoroughly mixed together. Mix only a small portion at a time as the paste sets hard within five to fifteen minutes depending on the amount of hardener used.

6. Smooth on the filler with a knife or stiff plastic to the shape of the damaged portion and allow to thoroughly dry a process which takes about six hours. After the filler has dried it is likely that it will have contracted slightly so spread on a second layer of filler if necessary.

7. Smooth down the filler with fine wet and dry paper wrapped round a suitable block of wood and continue until the whole area is perfectly smooth and it is impossible to feel where the filler joins the rest of the paintwork.

8. Spray on from an aerosol can, or with a spray gun, an anti-rust undercoat, smooth down with wet and dry paper, and then spray on two coats of the final finishing using a circular motion.

9. When thoroughly dry polish the whole area with a fine cutting paste to smooth the resprayed area into the remainder of the wing and to remove the small particles of spray paint which will have settled round the area.

10. This will leave the wing looking perfect with not a trace of the previous unsightly dent.

6. MAJOR CHASSIS & BODY REPAIRS

1. Major chassis and body repair work cannot be successfully undertaken by the average owner. Work of this nature should be entrusted to a competent body repair specialist who should have the necessary jigs, welding and hydraulic straightening equipment as well as skilled panel beaters to ensure a proper job is done.

2. If the damage is severe it is vital that on completion of repair the chassis is in correct alignment. Less severe damage may also have twisted or distorted the chassis although this may not be visible immediately. It is therefore always best on completion of repair to check for twist and squareness to ensure that all is correct.

3. To check for twist position the car on a clean level floor, place a jack under each jacking point and raise the car and take off the wheels. Raise or lower the jacks until points 'A' and 'E' are at the correct level (see Fig. 12.1). The datum line should now be 20 in. above the floor. With the points 'E' correctly set up, should it prove impossible to get the height of points 'A' the same, then the chassis is twisted, the amount of twist representing the difference in height of points 'A'.

4. After checking for twist check for squareness by taking a series of measurements on the floor. Drop a plumb line and bob weight from the lettered points on the chassis frame (Fig. 12.2.) to the floor and mark these points with chalk, and letter them to correspond with the letters in Fig. 12.2. Draw a straight line between each pair of letters and measure and mark the middle of each line. A line drawn on the floor starting at the middle of line 'A' and finishing at the middle of line 'E' should be quite straight and pass through the centres of the other lines. Diagonal measurements can also be made as a further check for squareness.

BODYWORK AND UNDERFRAME

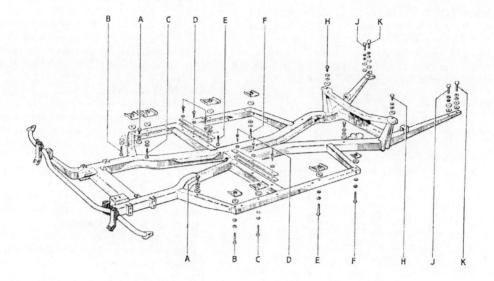

Fig. 12.1. THE BODY MOUNTING POINTS FOR HERALD MODELS
When testing for twist Points 'A' should be 25.5 in. (64.8 cm.) above the ground and at the same time Points 'E' 24.9 in. (63.3 cm.) above the ground.

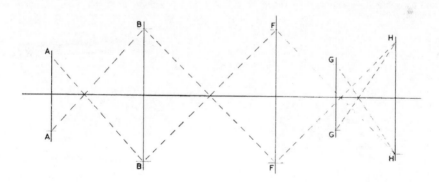

Fig. 12.2. The order in which to check the Herald chassis for twist and squareness.

187

CHAPTER TWELVE

7. **MAINTENANCE - HINGES & LOCKS**

Once every six months or 6,000 miles the door, bonnet and boot hinges should be oiled with a few drops of engine oil from an oil can. The door striker plates can be given a thin smear of grease to reduce wear and ensure free movement.

8. **DOOR LOCKS - REMOVAL & REPLACEMENT**

1. Close the window and remove the two interior handles from the door by pressing the escutcheon for each handle firmly against the trim panel, and pushing out the retaining pin with a short length of stiff wire or a thin electrical screwdriver.

Fig. 12.3. Layout of the Herald door after the trim panel has been removed.

2. The trim panel is held to the door by clips. Insert a screwdriver between the panel and the door and carefully prise the panel off one clip at a time. Take care not to damage the paintwork or the trim. The coil springs can now be taken off the window winder and door handle spindles.

3. With the aid of a thin electrical screwdriver take off the small spring clip from the remote control link attachment so as to release the link arm from the lock. Undo the screws holding the lock to the door. Press the lock in and down and remove it from the large aperture in the door. NOTE: On certain models it may also be necessary to remove the rear glass run channel before the lock can be removed. In this case, take off the rubber grommet from the top of the door rear face and undo the exposed bolt and washer. Undo the two bolts at the rear end of the inside face of the door and then remove the channel.

4. On replacement which is a straightforward reversal of the removal sequence, ensure that the escutcheon springs are replaced with the large coil ends against the door frame and the small coil ends bearing against the inner face of the escutcheon.

9. **STRIKER PLATE - REMOVAL, REPLACEMENT & ADJUSTMENT**

1. If it is wished to renew a worn striker plate mark its position on the door pillar so a new plate can be fitted in the same position.

2. To remove the plate simply undo the three Phillips screws which hold the plate in position. Replacement is equally straightforward.

3. To adjust the striker plate close the door and then push it hard against its sealing rubber. The door edge furthest from the hinges should move in very slightly.

4. Loosen the door striker plate screws and adjust the plate until the clearance is correct. Tighten the screws and check that the door closes properly without lifting or dropping, and that on the road it does not rattle.

10. **DOOR RATTLES - TRACING & RECTIFICATION**

1. The commonest cause of door rattle is a misaligned, loose, or worn striker plate but other causes may be:-

a) Loose door handles, window winder handles or

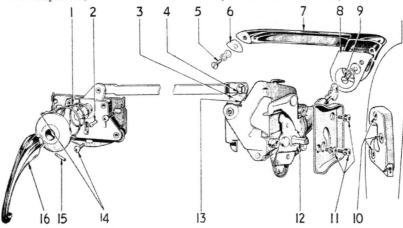

Fig. 12.4. THE DOOR LOCK AND REMOTE CONTROL MECHANISM

1 Escutcheon spring. 2 Split pin. 3 Remote control attachment. 4 Clip. 5 Screw. 6 Washer. 7 Exterior handle. 8 Washer. 9 Screw. 10 Phillips screws (striker plate). 11 Phillips screws (door). 12 Square shaft. 13 Lock frame. 14 Phillips screws. 15 Pin. 16 Interior handle.

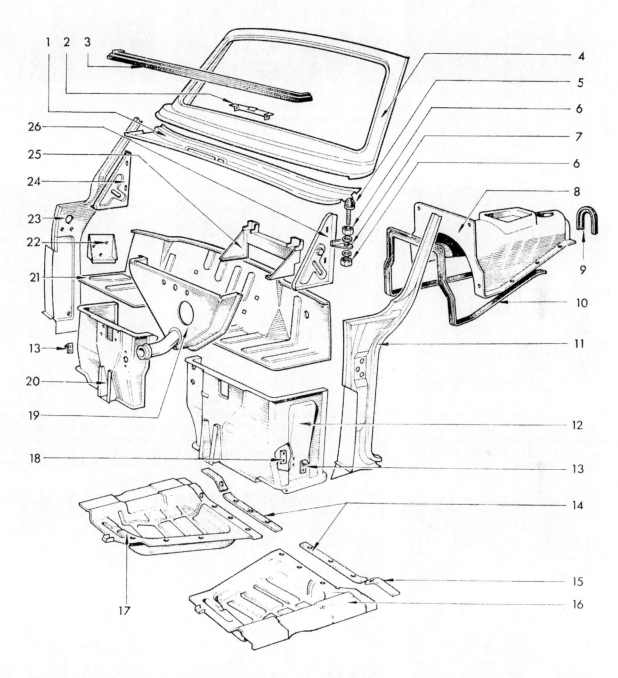

Fig. 12.5. EXPLODED VIEW OF THE CENTRE SECTION OF THE HERALD
1 Front deck. 2 Bracket—demister support. 3 Rubber seal. 4 Panel—screen surround. 5 Bonnet locator pin. 6 Lock nut. 7 Mounting plate. 8 Nut. 9 Seal—gearbox cover. 10 Seal—gearbox cover. 11 'A' post panel. 12 Panel—dash side. 13 Bonnet catch plate. 14 Sealing strip. 15 Sealing strip. 16 Front floor L.H. 17 Front floor R.H. 18 Stiffener plate. 19 Air box panel. 20 Panel—dash side. 21 Dash shelf panel. 22 Bracket—wiper motor. 23 'A' post panel. 24 Gusset panel. 25 Bracket assembly—battery. 26 Gusset panel.

189

This photographic sequence shows the steps taken to repair the dent and paintwork damage shown above. In general, the procedure for repairing a hole will be similar; where there are substantial differences, the procedure is clearly described and shown in a separate photograph.

First remove any trim around the dent, then hammer out the dent where access is possible. This will minimise filling. Here, after the large dent has been hammered out, the damaged area is being made slightly concave.

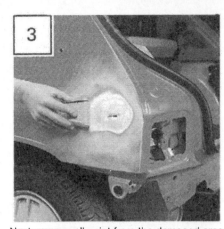

Next, remove all paint from the damaged area by rubbing with course abrasive paper or using a power drill fitted with a wire brush or abrasive pad. 'Feather' the edge of the boundary with good paintwork using a finer grade of abrasive paper.

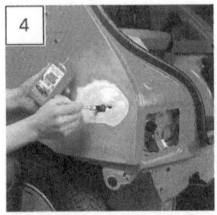

Where there are holes or other damage, the sheet metal should be cut away before proceeding further. The damaged area and any signs of rust should be treated with Turtle Wax Hi-Tech Rust Eater, which will also inhibit further rust formation.

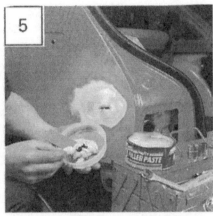

For a large dent or hole mix Holts Body Plus Resin and Hardener according to the manufacturer's instructions and apply around the edge of the repair. Press Glass Fibre Matting over the repair area and leave for 20-30 minutes to harden. Then ...

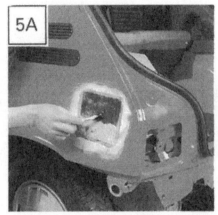

... brush more Holts Body Plus Resin and Hardener onto the matting and leave to harden. Repeat the sequence with two or three layers of matting, checking that the final layer is lower than the surrounding area. Apply Holts Body Plus Filler Paste as shown in Step 5B.

For a medium dent, mix Holts Body Plus Filler Paste and Hardener according to the manufacturer's instructions and apply it with a flexible applicator. Apply thin layers of filler at 20-minute intervals, until the filler surface is slightly proud of the surrounding bodywork.

For small dents and scratches use Holts No Mix Filler Paste straight from the tube. Apply it according to the instructions in thin layers, using the spatula provided. It will harden in minutes if applied outdoors and may then be used as its own knifing putting.

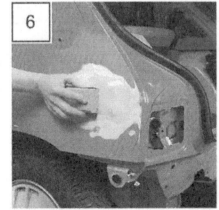

Use a plane or file for initial shaping. Then, using progressively finer grades of wet-and-dry paper, wrapped around a sanding block, and copious amounts of clean water, rub down the filler until glass smooth. 'Feather' the edges of adjoining paintwork.

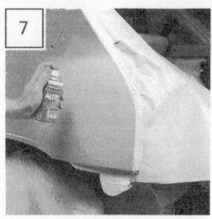

Protect adjoining areas before spraying the whole repair area and at least one inch of the surrounding sound paintwork with Holts Dupli-Color primer.

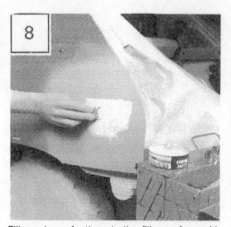

Fill any imperfections in the filler surface with a small amount of Holts Body Plus Knifing Putty. Using plenty of clean water, rub down the surface with a fine grade wet-and-dry paper - 400 grade is recommended - until it is really smooth.

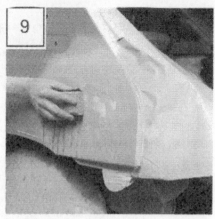

Carefully fill any remaining imperfections with knifing putty before applying the last coat of primer. Then rub down the surface with Holts Body Rubbing Compound to ensure a really smooth surface.

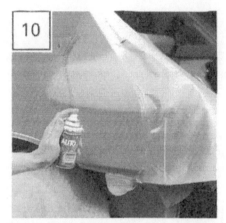

Protect surrounding areas from overspray before applying the topcoat in several thin layers. Agitate Holts Dupli-Color aerosol thoroughly. Start at the repair centre, spraying outwards with a side-to-side motion.

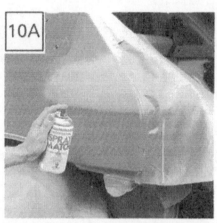

If the exact colour is not available off the shelf, local Holts Professional Spraymatch Centres will custom fill an aerosol to match perfectly.

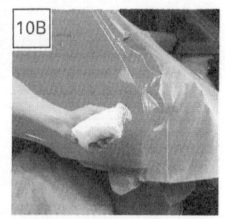

To identify whether a lacquer finish is required, rub a painted unrepaired part of the body with wax and a clean cloth.

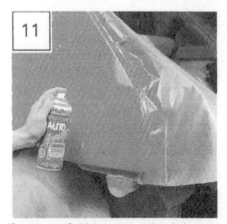

If *no* traces of paint appear on the cloth, spray Holts Dupli-Color clear lacquer over the repaired area to achieve the correct gloss level.

The paint will take about two weeks to harden fully. After this time it can be 'cut' with a mild cutting compound such as Turtle Wax Minute Cut prior to polishing with a final coating of Turtle Wax Extra.

When carrying out bodywork repairs, remember that the quality of the finished job is proportional to the time and effort expended.

CHAPTER TWELVE

door hinges.

b) Loose, worn or misaligned door lock components.

c) Loose or worn remote control mechanism.

2. It is quite possible for door rattles to be the result of a combination of the above faults, so a careful examination must be made to determine the causes of the fault.

3. If the nose of the striker plate is worn and as a result the door rattles, renew and then adjust the plate as described in Section 9.

4. Should the inner door handle rattle this is easily cured by fitting a rubber washer between the escutcheon and the handle.

5. If the nose of the door lock wedge is badly worn and the door rattles as a result, then fit a new lock as described in Section 8.

6. Should the hinges be badly worn, then they must be replaced.

11. WINDSCREEN GLASS - REMOVAL & REPLACEMENT

1. If you are unfortunate enough to have a windscreen shatter, fitting a replacement windscreen is one of the few jobs which the average owner is advised to leave to a professional mechanic. The owner who wishes to do the job himself will need the help of a friend.

2. Take off the windscreen wiper arms and blades, and remove the bright metal windscreen surround trims. The ends of the trims will be exposed by sliding the escutcheon at the middle of the bottom portion of the moulding to one side. The mouldings and escutcheon should then be carefully eased out of the rubber channel.

3. Work all round and under the outside edge of the windscreen sealing rubber with a screwdriver to break the Seelastik seal between the rubber and the windscreen frame flange.

4. Spread a blanket over the bonnet and with a friend steadying the glass from the outside sit in the passenger's seat and press out the glass and rubber surround with one foot. Place rag between your foot and the glass. The glass is started most easily at one of the corners.

5. Clean off all the old seelastik from the windscreen flange with petrol.

6. The windscreen is replaced after having smeared Seelastik sealing compound on the outside edge of the glass where it is covered by the rubber surround.

7. With the rubber surround and finisher fitted to the windscreen to that the joint is at the bottom insert 16 ft. of cord all round the channel in the rubber which will sit over the windscreen aperture flange. Allow the two free ends of the cord to overlap slightly.

8. Fit the windscreen from outside the car and with an assistant pressing the rubber surround hard against the body flange, slowly pull one end of the cord out moving round the windscreen and so drawing the lip of the rubber over the windscreen flange on the body.

12. INSTRUMENTS & SWITCHES - REMOVAL & REPLACEMENT

1. There are several small differences between the various models as far as the instruments and switches are concerned so not all of the methods

Fig. 12.6. EXPLODED VIEW OF DOOR, LOCK, HANDLES AND WINDOW REGULATOR MECHANISM

1 Metal door. 2 Door hinge assembly. 3 Hinge pin. 4 Sealing strip. 5 Inner sealing strip. 6 Quarter light outer frame assembly. 7 Top pivot bracket. 8 Thick washer. 9 Thin washer. 10 Semi tubular rivet. 11 Catch plate. 12 Glass channel. 13 Weatherstrip. 14 Quarter light inner frame. 15 Inner top pivot bracket. 16 Bottom pivot shaft assembly. 17 Bracket. 18 Spring. 19 Quarter light handle. 20 Push button. 21 Quarter light glass. 22 Glazing strip. 23 Spacer. 24 Washer. 25 Spring. 26 Tab washer. 27 Nut. 28 Bracket. 29 Lower bracket. 30 Door glass. 31 Regulator door glass channel. 32 Glazing channel strip. 33 Rear glass channel. 34 Weather curtain. 35 Tie rod. 36 Tie rod attachment clip. 37 Window regulator (winder) assembly. 38 Stud. 39 Leather washer. 40 Reinforcement plate. 41 Regulator mounting pivot. 42 Special washer. 43 Plain washer. 44 Lock washer. 45 Nut. 46 Window regulator handle. 47 Spring. 48 Escutcheon. 49 Handle fixing pin. 51 Lock assembly. 52 Dovetail plate. 53 Striker assembly. 54 Rubber seal. 55 Remote control assembly. 56 Waved washer. 57 Plain washer. 58 Clip. 59 Door handle (interior). 60 Spring. 61 Escutcheon. 62 Pin. 63 Exterior door handle. 64 Large seating washer. 65 Small washer. 66 Check link assembly. 67 Weather curtain. 68 Lock weather curtain. 69 Door trim panel. 70 Clip. 71 Pull handle.

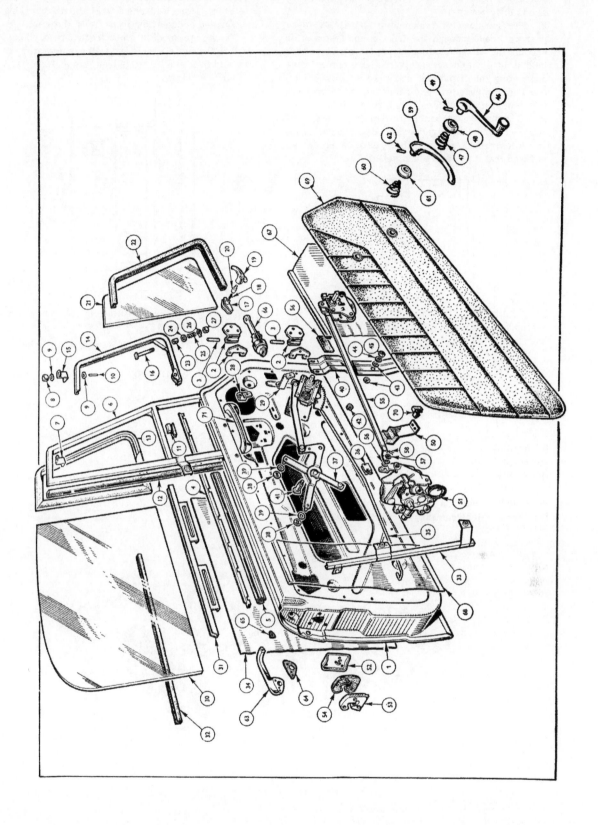

CHAPTER TWELVE

described here will apply to every car, but every car is fully covered.

2. The speedometer/tachometer is removed by pushing it out through the front of the fascia after undoing the two knurled nuts from the clamps which hold it in place; unscrewing the drive cable; disconnecting the trip reset cable (speedometer only) and Lucar connectors; and removing the instrument bulb holders. Replacement is a reversal of the removal sequence but watch two points. Ensure the rubber ring is undamaged and is positioned adjacent to the instrument rim and that where an earth lead is fitted that it is securely attached.

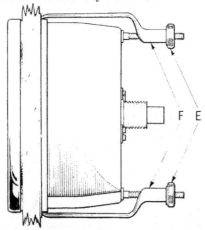

Fig. 12.7. Speedometer/tachometer instruments are held to the fascia panels by means of the brackets 'F' and knurled nuts 'E'.

3. The temperature/fuel gauge is removed by first pulling off the wires at the Lucar connectors and then the bulb holder from the back of the gauge. Undo the knurled nuts, spring washers, clamps and the earth lead. Push the gauge out through the front of the fascia. Replacement is a straightforward reversal of the removal sequence. Do not forget to refit the earth lead and check that the rubber ring is undamaged and seating properly on the rim.

4. On some models electrically operated gauges such as the fuel contents and water temperature units are held by a single knurled nut, spring washer, and 'U' shaped clamp.

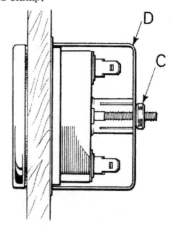

Fig. 12.8. On some models instruments are held in place by a bracket 'D' and knurled nut 'C'.

5. Switches of several different types may be fitted.

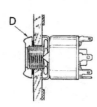

Fig. 12.9. Most switches are held in place by a bezel 'D' shown here on the ignition switch.

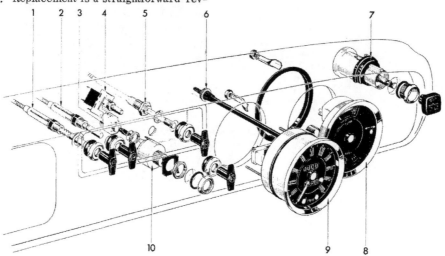

Fig. 12.10. THE SWITCHES, INSTRUMENTS AND CONTROLS USED ON THE 13/60 HERALD
1. Heater control cable. 2. Air distribution control. 3. Heat blower switch. 4. Lighting switch. 5. Choke control. 6. Trip cancelling control. 7. Windscreen washer wiper control. 8. Temperature fuel instrument. 9. Speedometer. 10. Ignition/starter switch.

194

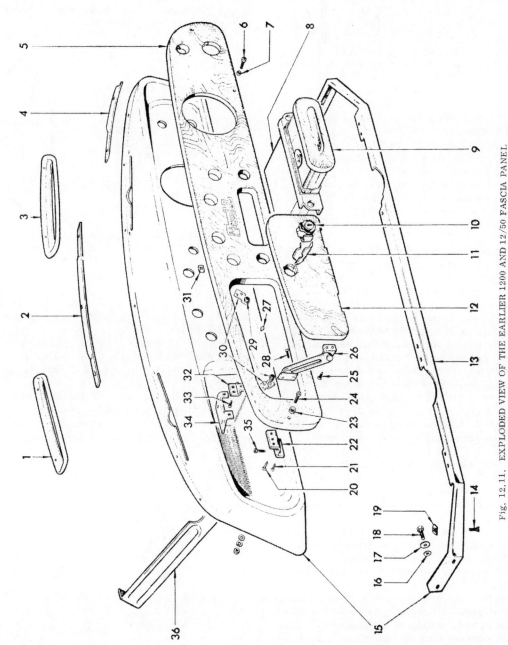

Fig. 12.11. EXPLODED VIEW OF THE EARLIER 1200 AND 12/50 FASCIA PANEL.

1 Finisher—demister vent. 2 Finisher—top edge—centre. 3 Finisher—demister vent. 4 Finisher—top edge—L.H. 5 Veneered panel. 6 Screw—panel attachment. 7 Cup washer—panel attachment. 8 Ash tray—support bracket. 9 Ash tray. 10 Glove box lock. 11 Finger pull. 12 Glove box lid. 13 Fascia rail. 14 Screw—fascia to rail. 15 Fascia panel assembly. 16 Washer. 17 Washer. 18 Bolt. 19 Fix nut. 20 Screw—hinge to lid. 21 Screw—hinge to lid. 22 Glove box—hinge. 23 Cup washer. 24 Screw. 25 Screw—link attachment. 26 Check link. 27 Screw. 28 Screw—link attachment. 29 Screw—link attachment. 30 Buffer bracket. 31 Clip. 32 Striker bracket. 33 Screw—striker bracket. 34 Tie bracket. 35 Screw—hinge to panel. 36 Bracket—glove box support.

195

CHAPTER TWELVE

The knobs on most of the switches are removed after depressing with a pin the spring loaded plunger holding the knob in place. The bezel holding the switch in place is then undone with the aid of a small screwdriver and the switch removed from behind the fascia panel. Note the position of the different coloured wires on the switch and pull off the wires with their Lucar connectors. Flick switches and the ignition switch are simply removed by disconnecting the wires at their backs and undoing the retaining bezel.

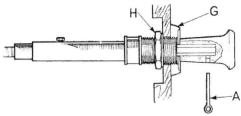

Fig. 12.12. Where push/pull switches with knobs are used the plunger in the shank must be depressed with a pin 'A' before the knob can be removed. Then undo the bezel 'G'.

13. WINDOW REGULATOR MECHANISM - REMOVAL & REPLACEMENT

1. If the window glass is very difficult to wind up and down try oiling the joints in the mechanism and the slots in the bottom channel assembly in which the regulator arms run before removing the mechanism.
2. If the mechanism is badly worn it will have to be replaced. To remove it from the car first remove the interior handles and trim. (See section 8 paras. 1 and 2).
3. Then free the two regulator arms from their slots in the bottom channel assembly by pulling off the small spring clips and leather washers and disconnecting the interconnecting link. Temporarily replace the window handle and raise the window to the fully closed position. It may be necessary to assist the window to rise by pulling it up manually. Wedge the glass in its highest position and then undo the four bolts holding the regulator mechanism and the three bolts which hold the regulator pivot plate in position and carefully manoeuvre the complete assembly out through the large hole in the inner panel of the door.
4. On replacement do not tighten down the seven bolts which hold the regulator mechanism in place until all the other connections have been made.

14. EXTERNAL DOOR HANDLES - REMOVAL & REPLACEMENT

1. Close the window and remove the two interior handles and the trim as described in section 8 paras. 1 and 2.
2. Undo the two small bolts which hold the door handle in place. One bolt is positioned on the door outside flange adjacent to the dovetail plate, and the other bolt is situated inside the door panel at the front end of the handle.
3. The handle can now be removed from the car. The pushbutton can be adjusted by loosening the locknut on the circular portion of the handle which fits inside the door and screwing the locknut in or out. When adjustment is correct, with the handle fitted, there should be $1/16$ in. clearance between the bolt head and the lock lever.
4. Replacement is a straightforward reversal of the removal sequence.

15. BONNET ASSEMBLY - REMOVAL, REPLACEMENT & ADJUSTMENT

1. Remove the earth lead from the battery and separate the wires for the lights and horns at their snap connectors located under the front of the bonnet and above the top centre of the grille.
2. Undo the three bolts (on later models only two) which holds each front over-rider in place and remove the over-riders. Take special note of the distance tube fitted over the top bolt on 1200 models.
3. Undo the nut and bolt which holds the bonnet support stay to the bracket on top of the suspension mounting, remove the two hinge bolts (5 and 6 in Fig. 12.13.) and lift off the bonnet.

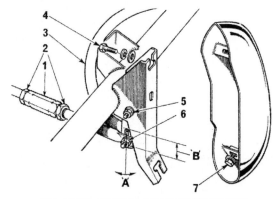

Fig. 12.13. BONNET ADJUSTMENT POINTS
1 Sleeve nut. 2 Locknuts. 3 Front tube. 4 Over-rider bolt.
5 & 6 Hinge bolts. 7 Over-rider bolt.

4. When refitting the bonnet adjust it so that a parallel clearance of $3/16$ in. (5 mm.) exists between the scuttle panel and the bonnet and make sure that the bonnet is on square and is not too high or low at its trailing edge.
5. To adjust the bonnet for parallel clearance at the scuttle loosen the locknuts on the arm (2 in Fig. 12.13) and then turn the large sleeve nut (1) until the clearance is correct. Tighten down the locknuts.
6. To adjust the height of the bonnet at the rear loosen the two nuts on the top bonnet stop fixed to the side of the scuttle and raise or lower the stop as required. Alternatively loosen the locknut which holds

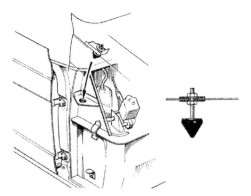

Fig. 12.14. The height of the bonnet is determined by the position of the adjustable rubber cone.

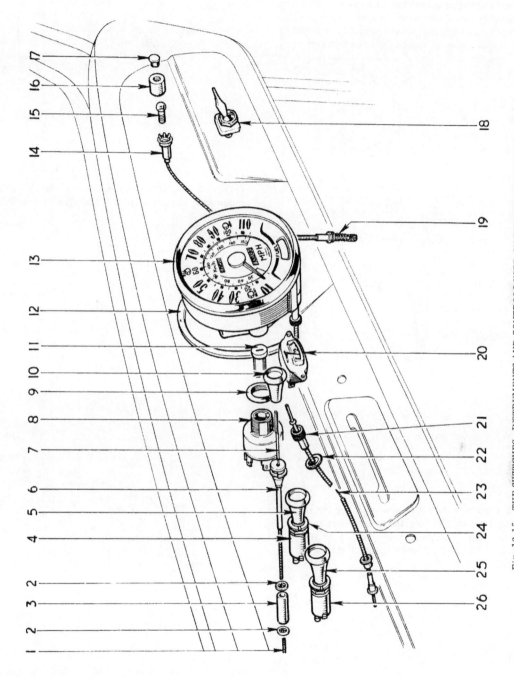

Fig. 12.15. THE SWITCHES, INSTRUMENTS AND CONTROLS USED ON THE HERALD 1200 AND 12/50

1 Choke control outer cable. 2 Clip. 3 Sleeve. 4 Switch. 5 Knob. 6 Choke control outer cable. 7 Choke control inner cable. 8 Starter ignition switch. 9 Bezel. 10 Knob. 11 Lock barrel. 12 Reinforcement ring. 13 Speedometer. 14 Bulb holder. 15 Bulb. 16 Lamp housing. 17 Lens. 18 Blower switch. 19 Trip cancelling cable. 20 Fuel gauge. 21 Speedometer drive outer cable. 22 Grommet. 23 Speedometer drive inner cable. 24 Bezel. 25 Knob. 26 Switch.

CHAPTER TWELVE

the cone shaped buffer to the bonnet and screw the buffer in or out to raise or lower the trailing edge. In both cases, then reset the bonnet catch plate at the bottom of the dash side panel so that the bonnet catches work correctly. The bolts at the front of the bonnet run in small slots so, here too, it is possible to make adjustments.

16. BUMPERS - REMOVAL & REPLACEMENT

1. The bumpers, front and rear, are part of the front and rear valances to which they are spot welded. The bumper rubbers are held in place by a flange and by cover plates at the end of each bumper.

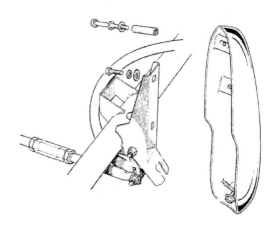

Fig. 12.18. The bolts holding the over-rider and bonnet in place.

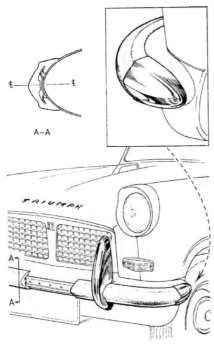

Fig. 12.16. The bumper rubber is held in place by a flange over which it fits and a cover plate (arrowed) at each end of the bumper.

3. At the rear the valance is split into three parts comprising the main bumper and wrap around bumpers and valances.

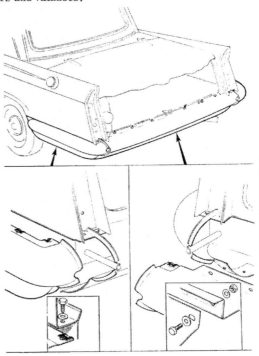

Fig. 12.19. The main and wrap around rear bumpers are part of the rear valances and are held in place as shown.

2. To remove a front valance open the bonnet and take out the bolts holding the over-riders in place. Undo the four Phillips screws, and the four bolts which hold the front valance in place and take off the valance.

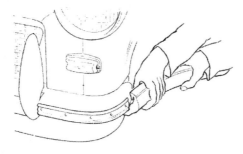

Fig. 12.17. Fitting a new bumper rubber in place.

4. To remove the main bumper take out the lens from the stop/rear lights and undo the bolts holding the over-riders in place. Undo the two screws which hold the boot lock striker plate in place in the middle of the flange on top of the bumper. Undo all the small bolts including those in the rear light compartments and lift off the valance/bumper assembly.

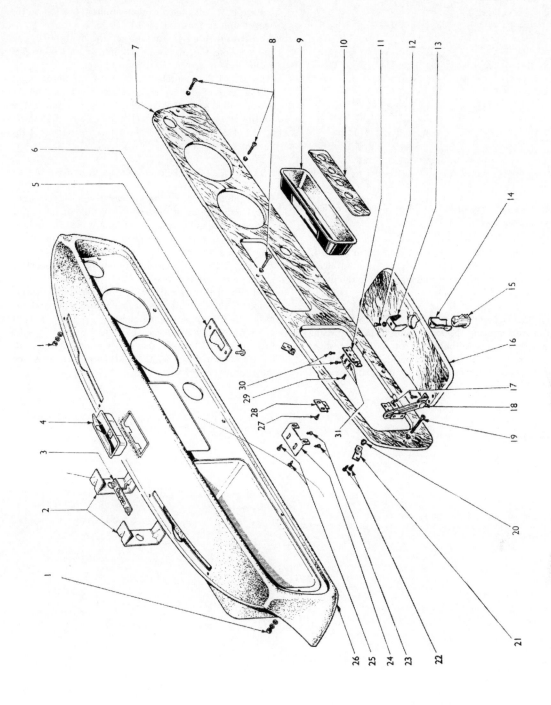

Fig. 12.20. EXPLODED VIEW OF THE HERALD 13/60 FASCIA PANEL

1 Nut—panel attachment. 2 Switch panel—saddle bracket. 3 Fascia—light switch. 4 Ash tray. 5 Light switch—cover plate. 6 Nylon stud—light switch. 7 Veneered panel. 8 Screw panel—attachment. 9 Switch panel. 10 Finisher plate. 11 Hinge—glove box. 12 Screw—lock clamp. 13 Lock clamp. 14 Finger pull. 15 Glove box lock. 16 Glove box lid. 17 Screw—link attachment. 18 Check link. 19 Screw—fascia attachment. 20 Rubber—buffer. 21 Buffer—bracket. 22 Screw—bracket attachment. 23 Screw—tie bracket. 24 Tie bracket. 25 Screw tie bracket. 26 Trimmed fascia. 27 Screw—striker bracket. 28 Striker bracket. 29 Screw—hinge to lid. 30 Screw—hinge to panel. 31 Screw—hinge to lid.

199

CHAPTER TWELVE

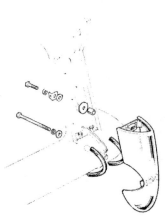

Fig. 12.21. Two bolts are used to hold each rear over-rider in place.

5. The wrap around bumpers are held in place in exactly the same way but to remove the left-hand wrap around bumper it is necessary to take out the fuel tank.

17. HEATER - REMOVAL & REPLACEMENT

1. Before removing the heater unit drain the cooling system, disconnect the battery, free the heater water hoses from the heater inlet and outlet pipes, and release the two electrical wires to the heater motor.

2. Disconnect the cable from the water control valve on the front of the heater unit and undo the screw from the 'L' shaped bracket holding the top of the heater unit to the dash panel.

3. From inside the car take off the dash millboard and release the air control cable from its bracket on the side of the air distribution box.

4. Pull off the demister tubes from the short pipes on the front of the air distribution box and undo the two nuts (one each side) holding the box in place. The heater unit is now free to be lifted from the bulkhead.

5. Replacement is a straightforward reversal of the removal sequence.

18. HEATER FAN MOTOR - REMOVAL & REPLACEMENT

1. If the fan motor fails to operate check the electrical connections at the switch and also at the motor before considering a suspect motor as the motor cannot be dismantled for repair, but must be exchanged for a new unit.

2. Disconnect the battery and free the two electrical wires to the heater motor.

3. Undo the three small bolts which hold the circular flange of the motor to the heater unit casing and remove the blower unit assembly.

4. Take off the brass nut from the threaded spindle in the centre of the impeller fan, and pull the fan off.

5. Replacement is a straightforward reversal of the removal sequence.

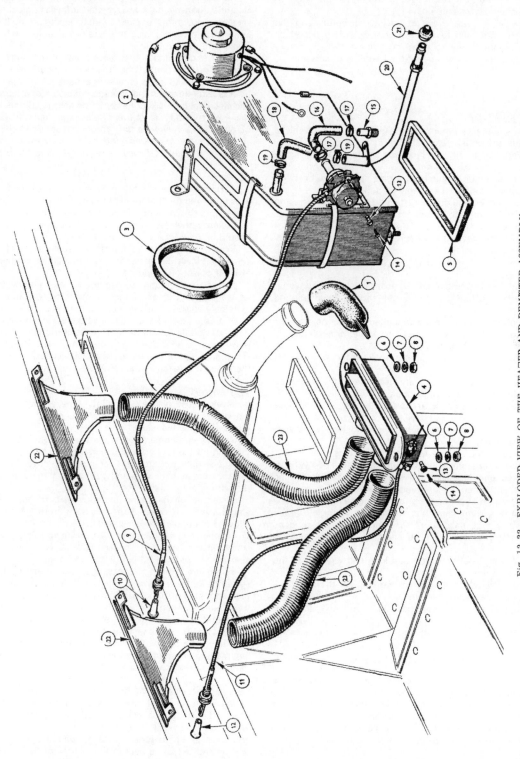

Fig. 12.22. EXPLODED VIEW OF THE HEATER AND DEMISTER ASSEMBLY

1 Valve. 2 Heater assembly. 3 Blower seal. 4 Outlet duct. 5 Gasket. 6 Plain washer. 7 Lock washer. 8 Hexagon nut. 9 Push/pull control (water valve). 10 Heat knob. 11 Push/pull control outlet duct. 12 Air distribution knob. 13 Trunnion. 14 Screw. 15 Toggle clamp. 16 Inlet hose. 17 Clip. 18 Outlet hose. 19 Clip. 20 Return pipe. 21 Adaptor. 22 Demister nozzle. 23 Demister hose.

201

Safety first!

Professional motor mechanics are trained in safe working procedures. However enthusiastic you may be about getting on with the job in hand, do take the time to ensure that your safety is not put at risk. A moment's lack of attention can result in an accident, as can failure to observe certain elementary precautions.

There will always be new ways of having accidents, and the following points do not pretend to be a comprehensive list of all dangers; they are intended rather to make you aware of the risks and to encourage a safety-conscious approach to all work you carry out on your vehicle.

Essential DOs and DON'Ts

DON'T rely on a single jack when working underneath the vehicle. Always use reliable additional means of support, such as axle stands, securely placed under a part of the vehicle that you know will not give way.

DON'T attempt to loosen or tighten high-torque nuts (e.g. wheel hub nuts) while the vehicle is on a jack; it may be pulled off.

DON'T start the engine without first ascertaining that the transmission is in neutral (or 'Park' where applicable) and the parking brake applied.

DON'T suddenly remove the filler cap from a hot cooling system – cover it with a cloth and release the pressure gradually first, or you may get scalded by escaping coolant.

DON'T attempt to drain oil until you are sure it has cooled sufficiently to avoid scalding you.

DON'T grasp any part of the engine, exhaust or catalytic converter without first ascertaining that it is sufficiently cool to avoid burning you.

DON'T allow brake fluid or antifreeze to contact vehicle paintwork.

DON'T syphon toxic liquids such as fuel, brake fluid or antifreeze by mouth, or allow them to remain on your skin.

DON'T inhale dust – it may be injurious to health (see *Asbestos* below).

DON'T allow any spilt oil or grease to remain on the floor – wipe it up straight away, before someone slips on it.

DON'T use ill-fitting spanners or other tools which may slip and cause injury.

DON'T attempt to lift a heavy component which may be beyond your capability – get assistance.

DON'T rush to finish a job, or take unverified short cuts.

DON'T allow children or animals in or around an unattended vehicle.

DO wear eye protection when using power tools such as drill, sander, bench grinder etc, and when working under the vehicle.

DO use a barrier cream on your hands prior to undertaking dirty jobs – it will protect your skin from infection as well as making the dirt easier to remove afterwards; but make sure your hands aren't left slippery. Note that long-term contact with used engine oil can be a health hazard.

DO keep loose clothing (cuffs, tie etc) and long hair well out of the way of moving mechanical parts.

DO remove rings, wristwatch etc, before working on the vehicle – especially the electrical system.

DO ensure that any lifting tackle used has a safe working load rating adequate for the job.

DO keep your work area tidy – it is only too easy to fall over articles left lying around.

DO get someone to check periodically that all is well, when working alone on the vehicle.

DO carry out work in a logical sequence and check that everything is correctly assembled and tightened afterwards.

DO remember that your vehicle's safety affects that of yourself and others. If in doubt on any point, get specialist advice.

IF, in spite of following these precautions, you are unfortunate enough to injure yourself, seek medical attention as soon as possible.

Asbestos

Certain friction, insulating, sealing, and other products – such as brake linings, brake bands, clutch linings, torque converters, gaskets, etc – contain asbestos. *Extreme care must be taken to avoid inhalation of dust from such products since it is hazardous to health.* If in doubt, assume that they *do* contain asbestos.

Fire

Remember at all times that petrol (gasoline) is highly flammable. Never smoke, or have any kind of naked flame around, when working on the vehicle. But the risk does not end there – a spark caused by an electrical short-circuit, by two metal surfaces contacting each other, by careless use of tools, or even by static electricity built up in your body under certain conditions, can ignite petrol vapour, which in a confined space is highly explosive.

Always disconnect the battery earth (ground) terminal before working on any part of the fuel or electrical system, and never risk spilling fuel on to a hot engine or exhaust.

It is recommended that a fire extinguisher of a type suitable for fuel and electrical fires is kept handy in the garage or workplace at all times. Never try to extinguish a fuel or electrical fire with water.

Note: *Any reference to a 'torch' appearing in this manual should always be taken to mean a hand-held battery-operated electric lamp or flashlight. It does NOT mean a welding/gas torch or blowlamp.*

Fumes

Certain fumes are highly toxic and can quickly cause unconsciousness and even death if inhaled to any extent. Petrol (gasoline) vapour comes into this category, as do the vapours from certain solvents such as trichloroethylene. Any draining or pouring of such volatile fluids should be done in a well ventilated area.

When using cleaning fluids and solvents, read the instructions carefully. Never use materials from unmarked containers – they may give off poisonous vapours.

Never run the engine of a motor vehicle in an enclosed space such as a garage. Exhaust fumes contain carbon monoxide which is extremely poisonous; if you need to run the engine, always do so in the open air or at least have the rear of the vehicle outside the workplace.

If you are fortunate enough to have the use of an inspection pit, never drain or pour petrol, and never run the engine, while the vehicle is standing over it; the fumes, being heavier than air, will concentrate in the pit with possibly lethal results.

The battery

Never cause a spark, or allow a naked light, near the vehicle's battery. It will normally be giving off a certain amount of hydrogen gas, which is highly explosive.

Always disconnect the battery earth (ground) terminal before working on the fuel or electrical systems.

If possible, loosen the filler plugs or cover when charging the battery from an external source. Do not charge at an excessive rate or the battery may burst.

Take care when topping up and when carrying the battery. The acid electrolyte, even when diluted, is very corrosive and should not be allowed to contact the eyes or skin.

If you ever need to prepare electrolyte yourself, always add the acid slowly to the water, and never the other way round. Protect against splashes by wearing rubber gloves and goggles.

When jump starting a car using a booster battery, for negative earth (ground) vehicles, connect the jump leads in the following sequence: First connect one jump lead between the positive (+) terminals of the two batteries. Then connect the other jump lead first to the negative (–) terminal of the booster battery, and then to a good earthing (ground) point on the vehicle to be started, at least 18 in (45 cm) from the battery if possible. Ensure that hands and jump leads are clear of any moving parts, and that the two vehicles do not touch. Disconnect the leads in the reverse order.

Mains electricity and electrical equipment

When using an electric power tool, inspection light etc, always ensure that the appliance is correctly connected to its plug and that, where necessary, it is properly earthed (grounded). Do not use such appliances in damp conditions and, again, beware of creating a spark or applying excessive heat in the vicinity of fuel or fuel vapour. Also ensure that the appliances meet the relevant national safety standards.

Ignition HT voltage

A severe electric shock can result from touching certain parts of the ignition system, such as the HT leads, when the engine is running or being cranked, particularly if components are damp or the insulation is defective. Where an electronic ignition system is fitted, the HT voltage is much higher and could prove fatal.

INDEX

A

A.C. Fuel pump - description - 63
A.C. Fuel pump - dismantling - 64
A.C. Fuel pump - examination & reassembly - 66
A.C. Fuel pump - removal & replacement - 64
A.C. Fuel pump - testing - 64
Air cleaners - removal, replacement, servicing - 63
Anti-freeze mixture - 58
Anti-roll bar - 172

B

Battery - charging - 146
Battery - electrolyte replenishment - 146
Battery - maintenance & inspection - 145
Battery - removal & replacement - 145
Big end bearings - examination & renovation - 36
Big end bearings - removal - 32
Body & chassis - maintenance - 185
Body repair sequence (colour) — 190/191
Bodywork & underframe - General description - 185
Bonnet assembly - removal, replacement & adjustment - 196
Brake calliper - removal, dismantling & reassembly - 138
Brake master cylinder - dismantling & reassembly - 134
Brake master cylinder - removal & replacement - 134
Brake pedal - removal & replacement - 136
Brakes - bleeding - 130
Brake seals - inspection & overhaul - 132
Braking system - fault finding chart - 141
Bulbs - 157
Bumpers — removal & replacement — 198

C

Camshaft & camshaft bearings - examination & renovation - 37
Camshaft removal - 32
Camshaft replacement - 42
Carburation - general description - 63
Carburetters - 66
Carpets - maintenance - 186
Clutch & actuating mechanism - general description - 94
Clutch & actuating mechanism - routine maintenance - 95
Clutch & actuating mechanism - specifications - 94
Clutch dismantling - 98
Clutch faults - 103
Clutch inspection - 98
Clutch judder - diagnosis & cure - 103
Clutch pedal - removal & replacement - 95
Clutch reassembly - 100
Clutch release bearing - adjustment - 102
Clutch release bearing - removal & replacement - 102
Clutch removal - 96
Clutch replacement - 96
Clutch slave cylinder - removal, dismantling, examination & reassembly - 100
Clutch spin - diagnosis & cure - 103
Clutch squeal - diagnosis & cure - 103
Clutch system - bleeding - 95
Coil springs - removal & replacement - 172
Condenser removal, testing & replacement - 87
Connecting rod - reassembly to crankshaft - 42
Connecting rod removal - 32
Contact breaker adjustment - 85
Contact breaker points - removing & replacing - 86
Control box - general description - 153
Cooling system - draining, flushing, filling - 55
Cooling system - fault finding chart - 61
Cooling system - general description - 54
Cooling system - specifications - 54
Crankshaft & main bearing removal - 34
Crankshaft examination & renovation - 36
Crankshaft rear seal - 44
Crankshaft replacement - 41
Crankshaft ventilation system - 34
Cut-out adjustment - 153
Cut-out & regulator contacts - maintenance - 153
Cylinder bores - examination & renovation - 36
Cylinder head - decarbonisation - 39
Cylinder head removal - 28
Cylinder head replacement - 46

D

Dampers - general description - 171
Dampers - specifications - 170
Damper units - removal & replacement - 172, 173
Decarbonisation - 39
Differential carrier assembly - removal & replacement - 125
Differential inner drive shafts - removal & replacement - 125
Differential unit - removal & replacement - 123
Disc & hub - removal & replacement - 138
Disc brake friction pad - inspection, removal & replacement - 136
Disc brakes - general description - 136
Disc brakes - maintenance - 136
Distributor & distributor drive replacement - 48
Distributor - dismantling - 88
Distributor drive removal - 32
Distributor - inspection & repair - 90
Distributor - lubrication - 87
Distributor - removal & replacement - 88
Distributor - reassembly - 90
Drive shaft universal joints - removal & replacement - 120
Door handles - removal & replacement - 194
Door locks - removal & replacement - 188
Door rattles - tracing & rectification - 188
Drum brakes - adjustment - 129
Drum brakes - general description - 129
Drum brake - shoe - inspection, removal & replacement - 130
Drum brakes - maintenance - 129
Dynamo - dismantling & inspection - 147
Dynamo - removal & replacement - 147
Dynamo - repair & reassembly - 148
Dynamo - testing in position - 146

E

Electrical system - fault finding chart - 166

INDEX

Electrical system - general description - 145
Electrolyte replenishment - 146
Engine - ancillary components - removing - 28
Engine - dismantling - 27
Engine - examination & renovation - 35
Engine - fault finding chart - 51
Engine - general description - 19
Engine removal - methods of - 20
Engine - reassembly - 40
Engine - replacement - 50
Engine - routine maintenance - 19
Engine - specifications & data - 14

F

Fan belt - adjustment - 58
Fan belt - removal & replacement - 60
Fault diagnosis – engine fails to start – 92
Fault diagnosis – engine misfires – 93
Fault finding chart - braking system - 141
Fault finding chart - cooling system - 61
Fault finding chart - electrical system - 166
Fault finding chart - engine - 51
Fault finding chart - fuel system & carburation - 83
Fault finding – ignition system – 92
Faults - clutch - 103
Fault symtoms – ignition system – 92
Flasher bulbs - 157
Flasher circuit - fault tracing & rectification - 154
Flexible hose - inspection, removal & replacement - 132
Flywheel replacement - 44
Flywheel starter ring - examination & renovation - 39
Front brake backplate - removal & replacement - 136
Front side & flasher bulbs - removal & replacement - 157
Front wheel cylinders - removal & replacement - 132
Front suspension - dismantling & reassembly - 174
Front wheel alignment - 178
Fuel contents gauge - fault tracing & rectification - 156
Fuel gauge - removal & replacement - 79
Fuel gauge sender unit - removal & replacement - 157
Fuel pump - description - 63
Fuel system & carburation - general description - 63
Fuel system & carburation - specifications - 62
Fuel tank - removal & replacement - 79
Fuse - Herald 13/60

G

Gearbox - dismantling - 108
Gearbox - examination & renovation - 110
Gearbox - general description - 105
Gearbox - reassembly - 112
Gearbox - removal & replacement - 105
Gearbox - specifications - 104
Gear change remote control - overhaul - 114
Gear selectors - removal & replacement - 114
Gudgeon pin - removal - 33

H

Handbrake adjustment - 135
Headlamp bulbs - removal & replacement - 157
Heater fan motor - removal & replacement - 200
Heater – removal & replacement – 200
Hinges & locks - maintenance - 188
Hoods - maintenance - 186
Horns - fault tracing & rectification - 156
Hydraulic system - bleeding - 130

I

Ignition system – fault finding – 92
Ignition system – fault symtoms – 92
Ignition system - general description - 85
Ignition system - specifications - 84
Ignition timing - 90
Input shaft - dismantling & reassembly - 110
Instrument panel & warning bulbs - removal & replacement - 158

L

Lubrication chart - 11

M

Main bearings - examination & renovation - 36
Main bearings - removal - 34
Mainshaft - dismantling & reassembly - 110
Maintenance - See Routine maintenance
Major chassis & body repairs - 186
Minor body repairs - 186

N

Number plate bulb - removal & replacement - 158

O

Oil filter - removal & replacement - 35
Oil pressure relief valve - 35
Oil pump - examination & renovation - 39
Oil pump - removal & dismantling - 35
Oil pump replacement - 44
Outer ball joint - removal & replacement - 182

P

Pinion oil seal - removal & replacement - 125
Piston & Connecting rod reassembly - 41
Pistons & piston rings - examination & renovation - 37
Piston replacement - 42
Piston ring removal - 32
Piston ring replacement - 42
Plugs & Leads – 92
Points - adjustment - 85
Propeller shaft & universal joints - specifications - 118
Propeller shaft - general description - 118
Propeller shaft - removal & replacement - 118

R

Rack & Pinion backlash - adjustment - 182
Rack & Pinion steering gear - 181
Radiator removal - 56
Rear axle - general description - 122
Rear axle - specifications - 122

INDEX

Rear drive shaft hubs - dismantling, examination & reassembly - 124
Rear drive shafts - removal & replacement - 123
Rear lamps - 158
Rear radius arms - removal & replacement - 173
Rear semi elliptic springs - removal & replacement 173
Rear wheel cylinders - removal & replacement - 134
Recommended lubricants - 10
Remote control assembly - overhaul - 114
Rocker arm adjustment - 48
Rocker assembly - dismantling - 31
Rockers & rocker shaft - examination & renovation - 38
Rocker shaft reassembly - 46
Routine maintenance - 7
Routine maintenance - engine - 19

S

Safety first! - 202
Servicing - See Routine maintenance
Side lamp bulbs - 157
Solex B28 21C-2 Carburetter - adjustment - 72
Solex B28 21C-2 Carburetter - description - 72
Solex B28 21C-2 Carburetter - dismantling & reassembly - 72
Solex B28 21C-2 Carburetter - removal & replacement - 72
Solex B30 PSE1 Carburetter - removal, dismantling & replacement - 74
Solex Carburetters - lack of fuel at engine - 80
Solex Carburetters - rich mixture - 80
Solex Carburetters - weak mixture - 80
Sparking plugs & leads - 92
Spark Plug Chart (colour) - 91
Starter motor bushes - inspection, removal & replacement - 153
Starter motor drive - general description - 152
Starter motor drive - removal & replacement - 152
Starter motor - dismantling & reassembly - 151
Starter motor - general description - 151
Starter motor - removal & replacement - 151
Starter motor - testing in engine - 151
Steering column - removal & replacement - 178
Steering column bushes - removal - 180
Steering gear - dismantling, reassembly & adjustment - 181
Steering gear - removal & replacement - 181
Steering - general description - 171
Steering - inspecting for wear - 171
Steering - specifications - 169
Steering wheel - removal & replacement - 178
Stop/rear & flasher bulbs - removal & replacement - 158
Strap drive - dismantling & reassembly - 121
Striker plate - removal, replacement & adjustment - 188
Stromberg 150 CD Carburetter - adjustments - 78
Stromberg 150 CD Carburetter - description - 76
Stromberg 150 CD Carburetter - dismantling & reassembly - 79
S.U. Carburetters - adjustment & tuning - 71
S.U. Carburetters - description - 66
S.U. Carburetters - dismantling - 69
S.U. Carburetters - examination & repair - 70

S.U. Carburetter float chamber - dismantling, examination & reassembly - 69
S.U. Carburetter float chamber - fuel level adjustment - 70
S.U. Carburetters - float chamber flooding - 71
S.U. Carburetters - float needle sticking - 71
S.U. Carburetters - piston sticking - 70
S.U. Carburetters - removal & replacement - 68
S.U. Carburetters - twin - synchronisation - 72
S.U. Carburetters - water and dirt in carburetter - 71
S.U. H1 Carburetter jet centring - 71
Sump - examination & renovation - 40
Sump, Piston, Connecting rod & big end bearing removal - 32
Sump replacement - 46
Suspension - front - dismantling & reassembly - 174
Suspension - general description - 171
Suspension - inspecting for wear - 171
Suspension - specifications - 169

T

Tail lamps - 158
Tappet reassembly - 46
Tappets - examination & renovation - 38
Temperature gauge & sender unit - removal & replacement - 58
Temperature gauge - fault finding - 58
Temperature gauge - fault tracing & rectification - 157
Thermostat - removal, testing, replacement - 56
Timing chain tensioner - removal & replacement - 38
Timing cover, gears & chain - removal - 32
Timing gears & chain - examination & renovation - 38
Timing gears - chain tensioner - cover replacement - 42
Timing - ignition - 90
Timing - valve - 42
Tonneau covers - maintenance - 186
Torque wrench settings - clutch & actuating mechanism - 94
Torque wrench settings - cooling system - 54
Torque wrench settings - fuel pump - 63
Torque wrench settings - gearbox - 104
Torque wrench settings - rear axle - 122
Torque wrench settings - suspension - steering - 170, 171
Trunnion housing bushes - removal & replacement - 173

U

Universal joints - dismantling - 120
Universal joints - general description - 118
Universal joints - inspection & repair - 120
Universal joints - reassembly - 121
Universal joints - specifications - 118
Upholstery - maintenance - 186

V

Valve adjustment - 48
Valves & valve seats - examination & renovation - 37
Valve & valve spring reassembly - 46
Valve guide removal - 31
Valve guides - examination & renovation - 40

INDEX

Valve removal - 31
Voltage regulator adjustment - 153

Water pump - dismantling & reassembly - 58
Water pump - removal & replacement - 57
Wheelbox - removal & replacement - 154
Window regulator mechanism - removal & replacement — 196

Windscreen wiper arms - removal & replacement - 154
Windscreen wiper mechanism — fault diagnosis & rectification - 154
Windscreen wiper mechanism - maintenance - 154
Windscreen wiper motor - dismantling, inspection & reassembly - 154
Wiper motor - removal & replacement - 154
Wiring diagrams - 159